Nice Numbers

John Barnes

Nice Numbers

 Birkhäuser

John Barnes
Caversham
England

ISBN 978-3-319-46830-3   ISBN 978-3-319-46831-0 (eBook)
DOI 10.1007/978-3-319-46831-0

Library of Congress Control Number: 2016957844

© Springer International Publishing Switzerland 2016
This work is subject to copyright. All rights are reserved by the Publisher, whether the whole or part of the material is concerned, specifically the rights of translation, reprinting, reuse of illustrations, recitation, broadcasting, reproduction on microfilms or in any other physical way, and transmission or information storage and retrieval, electronic adaptation, computer software, or by similar or dissimilar methodology now known or hereafter developed.
The use of general descriptive names, registered names, trademarks, service marks, etc. in this publication does not imply, even in the absence of a specific statement, that such names are exempt from the relevant protective laws and regulations and therefore free for general use.
The publisher, the authors and the editors are safe to assume that the advice and information in this book are believed to be true and accurate at the date of publication. Neither the publisher nor the authors or the editors give a warranty, express or implied, with respect to the material contained herein or for any errors or omissions that may have been made.

Printed on acid-free paper

This book is published under the trade name Birkhäuser
The registered company is Springer International Publishing AG
The registered company address is: Gewerbestrasse 11, 6330 Cham, Switzerland
(www.birkhauser-science.com)

To
Bobby,
Janet, Helen,
Christopher and Jonathan

Preface

This book is based on lectures given originally at Reading University and more recently at Oxford as part of the Continuing Education program of Oxford University in England. In a sense it is a sequel to *Gems of Geometry* (now in its second edition) and which is also based on lectures given at Reading and Oxford.

Having given the geometry lectures a few times, it was suggested that I should develop another course based on aspects of numbers. I was extremely busy at the time updating my book on the Ada programming language so settled on a short course of just five lectures entitled Nice and Noughty (sic) Numbers. However, as the time approached I was still too busy and so the course was kindly given by Aruna Hardy. The next year, the pressure of life abated and I was able to gave a full ten lecture course.

I have since given the course a number of times at both Reading and Oxford and the notes eventually matured into this book. The lectures divide into two kinds: some address the theme of simple number theoretic topics such as prime numbers and cryptography, whereas others address the theme of daily needs and pleasures such as keeping track of time and enjoying music.

There are ten basic lectures. The first lecture is entitled Measures. Measuring and counting various things are fundamental human activities. In earlier times one might have been interested in measuring the size of a field and counting cows; these days one might be more likely to measure the size of a garden and count money. The lecture starts by asking students what is their favourite number and why; this not only breaks the ice but lays the foundation for many topics such as what would be a good base for counting. That takes us into prime numbers, divisibility, factors, and perfect numbers; the lecture concludes with a survey of weights and measures and currencies.

The second lecture leads on from perfect numbers into Amicable numbers and provides a good opportunity to introduce modular arithmetic. This is followed by Probability and is more light-hearted and ends with applications to games such as craps and poker. We then return to the numerical theme by considering Fractions of various forms including the amazing Egyptian system of unit fractions. This is followed by a lecture on Time including the calendar and sunshine.

There are then two more lectures on the numerical theme covering Notations of various forms and Prime numbers and two lectures of a more fun nature covering Bell ringing and the evolution of Musical scales.

The final lecture looks at the topic of cryptography which has become so important with electronic communication. And for final light relief it concludes by looking at the gaits of animals and two popular puzzles.

There are also a number of appendices. Some provide additional material on topics such as Pascal's triangle and polydivisibility, whereas a few cover material from a short course called Puzzles and Pastimes which I have also given at Reading and Oxford. Thus the appendix on Groups draws on examples from Bell ringing and also leads into the amazing topic of Rubik's cube from the puzzles course. The final appendix considers the musical keyboards designed by Mersenne who is better known for his numbers; it concludes with a discussion of the Tonnetz schema for revealing harmony and the strange relation between music and the topology of the torus.

The main lectures contain some exercises (harder ones are marked with asterisks) but answers are not provided since I anticipate giving the course again.

An important issue when writing a book is to consider who might enjoy reading it. The mathematical background required is not hard (just a bit of simple algebra really) and is the kind of material anyone who studied a scientific subject to the age of perhaps 16 would have encountered. One group of potential readers is therefore young people with a zest for knowledge (I would have been delighted to have had such a book when I was 16). Another group as shown by students on courses includes those of maturer years who might like to know more about topics that they enjoyed when younger.

Students attending the courses are from various backgrounds – of all ages and sexes. Some have little technical background but revel in activities such as throwing dice, ringing handbells, and attempting to simulate the gaits of camels; others have serious technical experience and enjoy perhaps a nostalgic trip visiting some familiar topics and also meeting new ones.

I have made no attempt to avoid using mathematical notation wherever it is appropriate (readers can skip hard bits if they are weary). I am saddened that many popular mathematical books strive to avoid mathematics because some publisher once said that every time an equation is added, the sales divide by two. But I have aimed to provide various tables and illustrations to enliven the text.

I must now thank all those who have helped me in this task. First, a big thank you to my wife, Bobby, who helped with typesetting and photography, to my daughter Janet, who provided much background material, and to my daughter Helen for advice on design, and to David Godwin for information on the noble art of bell-ringing. Thanks also to the (anonymous) reviewers for their input and to Pascal Leroy who was a great help in finding a number of errors and suggesting many improvements and to Tucker Taft for assistance on analysing the performance of Ackermann's function and to Ahlan Marriott for the gift of a corkscrew with a message in base 4.

Thanks also to colleagues in Continuing Education at Oxford and especially to Aruna Hardy for digging me out of a hole by giving the first short numbers course and to Julian Gallop and Iryna Schlackow for some inspiring examples of probability and numerical puzzles.

I am grateful to many students on my courses for their feedback and encouragement. I must especially thank: Felix Lam for introducing me to the life of John William Colenso who wrote fascinating books on Arithmetic and became

Bishop of Natal; Susan Vaughan for help with the presentation of the strange material on music; Rita Sawrey-Woodwards for taking the tricky photographs of measuring devices and other artifacts; and Felicity Wood for introducing me to the Maria Theresa thaler. Susan Vaughan also introduced me to Geoff Chew at Royal Holloway who kindly brought me up to date on the world of music theory.

I am also grateful to the authors of the many books that I have read and enjoyed. I cannot mention them all here but I must mention a few. Two old favourites are *An Introduction to Probability Theory* by Feller and *On the Sensations of Tone* by Helmholtz. Regarding the ancient world, *Mathematics in the Time of the Pharaohs* by Richard Gillings is excellent. For more traditional number theory topics, I am grateful to my daughter for lending me her copies of two books entitled *Elementary Number Theory* by Kenneth Rosen and David Burton. Another wonderful book is *Makers of Mathematics* by Stuart Hollingdale.

For general fun, *The Penguin Dictionary of Curious and Interesting Numbers* by David Wells is vital and for puzzles the classic *Amusements in Mathematics* by Henry Dudeney is hard to beat. And I must thank David Singmaster for giving me a copy of his intriguing little book on the Rubik cube.

Probably the first book to trigger my interest in numbers was *Arithmetic* by D A Young. This was aimed at accountants in the old days when everything was worked out with a pencil (my father was an accountant). But among basic stuff were little advanced bits in small print which dabbled in odd topics such as finding a cube root, continued fractions, and recurring decimal fractions. I am sure I spent hours browsing through this when still in short trousers!

Finally, and most important of all, I must thank Dorothy Mazlum, Sabrina Hoecklin, and all others at Springer who made this book actually happen.

I hope that all those who read or browse through it will find something to enjoy. I enjoyed writing it and learnt a great deal in the process.

John Barnes
Caversham
England
September 2016

Contents

1 Measures   1
Favourite numbers, Prime numbers, Factors, Weights and measures, Musical notes, Currencies, Further reading, Exercises.

2 Amicable Numbers   25
Perfect numbers, Modular arithmetic, Mersenne the monk, Amicable numbers, Amicable multiplets, Sociable cycles, Fermat numbers, Fibonacci numbers, Further reading, Exercises.

3 Probability   43
Heads or tails, Distributions, Shake, rattle, and roll, The normal distribution, An abnormal distribution, Tossing for π, Double or quits, A medical problem, Which box has the prize?, Craps and poker, Further reading, Exercises.

4 Fractions   69
Real numbers, Vulgar fractions, Egyptian fractions, Dividing loaves, The table of $2/n$, The method of false position, Decimal fractions, Roots, Continued fractions, The eye of Horus, Further reading, Exercises.

5 Time   97
Basic rhythms, The Roman calendar, The Gregorian calendar, Start of the year and quarters, The week, Time of day, Sunshine, Further reading, Exercises.

6 Notations   115
Types of notation, Roman numbers, Babylonian system, Place systems and bases, Divisibility, Fractions and bases, Fermat's Little Theorem, Further reading, Exercises.

7 Bells — 137
Rounds and plain hunting, Plain Bob Minimus, Plain Bob Doubles, Stedman, Grandsire, Groups, Further reading, Exercises.

8 Primes — 161
Greatest Common Divisor, Prime factors, Fermat's method, Eratosthenes revisited, Complex numbers, Complex primes, Polynomials, Further reading, Exercises.

9 Music — 179
Frequency and vibrations, The Pythagorean scale, Cents, Just intonation, Minor scales, Meantone temperament, Equal temperament, Other arrangements, Dividing the octave, Further reading.

10 Finale — 205
The RSA algorithm, Linear congruences, Euler's function, Encryption and decryption, Code blocks, Animal gaits, Towers and rings, Further reading.

A Ackermann — 227
Recursion and iteration, Further reading.

B Pascal's Triangle — 231
Basic properties, Fibonacci numbers, Squares and pyramids, Further reading.

C Stochastics — 239
War games, Queuing, Further reading.

D Polydivisibility — 245
The problem, Other bases.

E Groups — 249
Basics, Subgroups, Generators, Cosets, Permutations, Polyhedral groups, Direct products, The quaternion group, Further reading.

F Rubik — 265
Basics, Restoring a cube, Summary of restoration, The cube group, Cosets, Finally, Further reading.

G	**Differences**	**287**
	Rings, Towers, Fibonacci numbers, Hats, Further reading.	
H	**Chinese Remainders**	**295**
	Linear congruences, Simultaneous equations, Pirates, Eggs, Squares.	
J	**Mersenne**	**303**
	Mersenne's 31-note keyboard, A schema, The 12-note keyboards, An 18-note keyboard, A 26-note keyboard, The 31-note keyboard, Intervals, A better keyboard, Modulation, Tonnetz revisited, Further reading.	

Bibliography	325
Index	327

1 Measures

THIS FIRST LECTURE looks at some simple properties of positive whole numbers such as whether they are prime and what factors they have.

The number of factors of a number is important in determining whether it is good as a basis for measuring and counting. Accordingly, we also look at some historical weights and measures and currencies as well as at musical intervals which are another source of curious measures.

Favourite numbers

MANY PEOPLE have favourite numbers. Some are said to have mystical significance, some might be based on birthdays, and of course we all need easy to remember pin numbers.

A recent survey of some twenty (a score) of numerate folk produced the following list of favourite numbers. Names have been changed for anonymity.

0   Three people chose zero. Julia chose it because it filled a gap in number theory. Jim chose it to be different (but it seems he was unsuccessful because it was one of the most popular numbers!) and also because it differs most from all the others and because it is useless outside mathematics (hmm) and most importantly it is beautiful. Jane chose zero because it shows she has no defects in her software and research.

1/3 Trevor chose this because calculators just can't quite get it right.

1   My school number said Richard. Also chosen for the same reason by Bob. Angela liked it for consistency.

$\sqrt{2}$ Liked by Cyril simply because he likes the square root sign.

2   The first choice of John A because two heads are better than one and it takes two to tango (which is fun) whereas three is a crowd. Note that tango is Latin for I touch. See also 12. Also chosen by John B because it's the smallest prime and the only even prime; moreover, for a computer scientist 2 makes life easier.

e Simon and William chose e becomes it pops up/crops up in odd/interesting places. See also 5040.

π Liked by Yorick and Dod because it underpins the order of the universe and goes around.

4   An old astrology book said it was my lucky number and I can hit it on a dartboard said Trevor.

2 Nice Numbers

7
Five people chose 7. Reasons were: it's magic (Joseph), don't know (Adrian) and, I seemed to have success when picking this number as a child in games of chance (John C – an early big time gambler clearly). Big enough to be interesting but small enough to remember says Daniel. Hmm, I wonder if Daniel has trouble with his pin numbers. Finally, Bertie hadn't really thought about it but it was a useful small prime and related to his favourite colour, Red. (I wonder what the relation is.)

12
Another choice of John A because it has more factors than 10. (I didn't know that we were allowed more than one choice.) We should all use base 12 he says. I agree and there is a base 12 society but I think it has a slim chance now.

13
A significant date for Derek. And not proven to be unlucky said Robert.

17
I'm not sure why I like it said Stewart.

22
This is guaranteed to come up whenever I play roulette said Robin.

23
Reason unknown said Peter.

27
A good age to be (it was a long time ago for me). But quite recent for the owner I think. Ladies should never reveal their age but Alice has given us a lower bound.

34
Is the favourite Fibonacci number of Rustin.

509
Ackermann(3, 6) – a good compiler test for Clive.

777
Looks nicer than 999 and not the number of the Beast (666) said Douglas.

998
Engine capacity of Malcolm's first car.

1729
Alfred chose this for the same reason as Ramanujan. It is the smallest number that can be expressed as the sum of two cubes in two different ways: $12^3 + 1^3 = 10^3 + 9^3$. This relates to a story about a taxi number. The taxi was taken by Hardy when visiting Ramanujan who was ill. Also chosen by Martin who thought it might be prime; but it is 19×91 which is rather nice.

1961
A good number for palindromes (invert the page) and Gerald won a bottle of whisky in a raffle with it.

5040
This was Simon's second choice (after e). 5040 is of course 7! which is the number of changes for a classical peal of bells.

77385107
Turn it upside down and it is my name says ... – well do it on your calculator and see who said that.

137438691328
This is the seventh perfect number. The choice of Tony, just to be awkward/different he says. And he succeeded in being different unlike Jim who chose zero.

It is interesting that 7 was the most popular. It seems that if you ask people to choose a number between 1 and 10 then most people choose 7. Indeed, seven has an air of mysticism about it. There are seven days in the week, seven deadly sins, seven wonders of the ancient world, seven Vestal Virgins, seven muses, and so on.

Moreover, in the world of entertainment we have the film Seven Brides for Seven Brothers, Snow White had seven dwarves and there were seven Von Trapp children in the Sound of Music. And the person going to St Ives encountered a man with seven wives, seven cats, seven kits et cetera.

And there are seven notes in an octave CDEFGAB although it is called an octave because if we count inclusively as the Romans did then we count the C at both ends. A similar oddity occurs with a fortnight being quinze jours in French.

I am not sure about my own favourite number. I have a liking for 37 because 111 is 3×37 and I find that surprising. And 111 is quite pretty. Why not choose 111 therefore? Because it isn't prime perchance. But maybe our real favourite should be 12 for many reasons that will emerge.

It is interesting that among this group of a score of people nobody chose any complex numbers or negative numbers. But pleasing to have transcendental numbers and a surd as well as integers and one rational fraction. Ackermann's function which gives 509 is explained in Appendix A.

Another similar group of about twenty people showed much the same sort of pattern. Again 7 was the most popular.

Prime numbers

A PRIME NUMBER p is a number that has no factors, that is no other number divides exactly into it (and so without a remainder) other than 1 and the number p itself.

The first few primes up to 120 are

    2,   3,   5,   7,
 11, 13, 17, 19,
 23, 29,
 31, 37,
 41, 43, 47,
 53, 59,
 61, 67,
 71, 73, 79,
 83, 89,
 97,
101, 103, 107, 109,
113

All primes are odd except for 2 and none end in 5 except for 5 itself. The distribution of primes otherwise seems to be quite chaotic – that is, there is no pattern at all. Thus among the first few shown above, they seem to be thinning out with there being only one prime in the decade from 90 to 100 but then there are suddenly four in the decade from 100 to 110.

4 Nice Numbers

Quite a lot of general things are known about the distribution of prime numbers. But there are few really specific facts known about them.

An important general fact is that prime numbers go on for ever. This was known to Euclid and he gave a proof which is Proposition 20 of Book IX of his Elements. His proof is a bit obscure in some ways because it is essentially done by geometric reasoning.

A modern description might be as follows. Suppose $p, q, r,$ are any primes. Then consider

$$x = pqr + 1$$

If x is prime then we have found a new prime. If x is not prime then it must have prime factors. None of p, q, r can be prime factors since they give a remainder of 1 on division. So there must be some other prime and so they go on for ever.

We can try this on the first few primes.

$2 \times 3 + 1 = 7,$
$2 \times 3 \times 5 + 1 = 31,$
$2 \times 3 \times 5 \times 7 + 1 = 211,$
$2 \times 3 \times 5 \times 7 \times 11 + 1 = 2311,$

7, 31, 211, and 2311 are all new primes. The next few are products of new primes

$2 \times 3 \times 5 \times 7 \times 11 \times 13 + 1 = 30{,}031 = 59 \times 509$
$2 \times 3 \times 5 \times 7 \times 11 \times 13 \times 17 + 1 = 510{,}511 = 19 \times 97 \times 277$
$2 \times 3 \times 5 \times 7 \times 11 \times 13 \times 17 \times 19 + 1 = 9{,}699{,}691 = 347 \times 27{,}953$
$2 \times 3 \times 5 \times 7 \times 11 \times 13 \times 17 \times 19 \times 23 + 1 =$
$\qquad 223{,}092{,}871 = 317 \times 703{,}763$
$2 \times 3 \times 5 \times 7 \times 11 \times 13 \times 17 \times 19 \times 23 \times 29 + 1 =$
$\qquad 6{,}469{,}693{,}231 = 331 \times 571 \times 34{,}231$

The next one is actually a prime (these primes are sometimes called Euclidean primes)

$2 \times 3 \times 5 \times 7 \times 11 \times 13 \times 17 \times 19 \times 23 \times 29 \times 31 + 1 =$
$\qquad 200{,}560{,}490{,}131$

But then more products of primes are produced

$2 \times 3 \times 5 \times 7 \times 11 \times 13 \times 17 \times 19 \times 23 \times 29 \times 31 \times 37 + 1 =$
$\qquad 7{,}420{,}738{,}134{,}811 = 181 \times 60{,}611 \times 676{,}421$
$2 \times 3 \times 5 \times 7 \times 11 \times 13 \times 17 \times 19 \times 23 \times 29 \times 31 \times 37 \times 41 + 1 =$
$\qquad 304{,}250{,}263{,}527{,}211 = 61 \times 450{,}451 \times 11{,}072{,}701$
$2 \times 3 \times 5 \times 7 \times 11 \times 13 \times 17 \times 19 \times 23 \times 29 \times 31 \times 37 \times 41 \times 43 + 1 =$
$\qquad 13{,}082{,}761{,}331{,}670{,}031 = 167 \times 78{,}339{,}888{,}213{,}593$

and so on. The next Euclidean prime occurs with the product up to 379 but is too large to list here.

One holy grail that has been sought is a formula that always produces primes. The famous Swiss mathematician Leonhard Euler (1707–1783) found that the expression $x^2 + x + 41$ is prime for all values of x from 0 to 39. Sadly it doesn't work for $x = 40$ since it becomes $1600+40+41 = 1681 = 41^2$. More recently it has been discovered that positive numbers generated by the formula

$$(k+2)\{1-[wz+h+j-q]^2 - [(gk+2g+k+1)(h+j)+h-z]^2 - [2n+p+q+z-e]^2 - [16(k+1)^3(k+2)(n+1)^2+1-f^2]^2 - [e^3(e+2)(a+1)^2+1-o^2]^2 - [(a^2-1)y^2+1-x^2]^2 - [16r^2y^4(a^2-1)+1-u^2]^2 - [((a+u^2(u^2-a))^2-1)(n+4dy)^2+1-(x+cu)^2]^2 - [n+l+v-y]^2 - [(a^2-1)l^2+1-m^2]^2 - [ai+k+1-l-i]^2 - [p+l(a-n-1)+b(2an+2a-n^2-2n-2)-m]^2 - [q+y(a-p-1)+s(2ap+2a-p^2-2p-2)-x]^2 - [z+pl(a-p)+t(2ap-p^2-1)-pm]^2\}$$

are always prime. Gosh! Moreover, every prime is generated by some choice of values for the 26 integers $a \ldots z$. A large problem is that most numbers generated by the formula are negative and so don't count! There is an interesting discussion on this topic at the website mentioned at the end of the lecture.

Although it is known that primes go on for ever, it is not known whether pairs two apart such as 71 and 73 also do but it seems almost certain. It is also conjectured that decades of four primes such as 101, 103, 107, 109 also go on for ever. There are many other conjectures such as that of Christian Goldbach (1690–1764) that every even number is the sum of two primes or a prime plus one. Note that we do not consider 1 to be a prime otherwise it would be more simply stated as that every even number is the sum of two primes.

Factors

IF A NUMBER has many factors then it is generally more useful for counting and measuring. It is interesting to calculate the sum of the factors of a number.

If the sum of the factors is less than the number itself then it is said to be deficient. 10 is an example of a deficient number. Its factors are 1, 2, and 5 and these add up to 8 which is less than 10. If the sum of the factors is more than the number then it is said to be abundant. 12 is an example of an abundant number. Its factors are 1, 2, 3, 4, and 6 and these add up to 16.

We can denote the sum of the factors of n by $s(n)$. The ratio $s(n)/n$ is a measure of its abundance. If the sum of the factors equals the number itself then it is said to be perfect. 6 is a perfect number since its factors are 1, 2, and 3 and these add up to 6.

A number is said to be superabundant if it is more abundant than all numbers less than it. Clearly, 12 which is the first abundant number is also superabundant. Perhaps 12 should really be our favourite number.

Note that the powers of 2 (16, 32, 64 etc.) are always slightly deficient although they approach the ratio of 1. So an abundant number must have factors other than 2.

6 Nice Numbers

On the other hand an abundant number need not be even. But it does need to have plenty of factors. The smallest abundant odd number is 945 which is 27.5.7 (that is $3^3 \times 5 \times 7$). Its factors are 1, 3, 5, 7, 9, 15, 21, 27, 35, 45, 63, 105, 135, 189, 315 which add up to 975 giving a ratio of 1.031....

(We sometimes use . rather than × to denote multiplication to save space. Context should distinguish it from a decimal point.)

The table opposite gives the factors of all numbers up to 30 and some selected numbers above.

The superabundant numbers also go on for ever with ever increasing ratio. The table below gives the first few and their prime factors. Note that 60 is especially superabundant in the sense that it is a great improvement over the previous superabundant number which is 48.

The first few factorials are superabundant. But 8! is not. (Remember that 8! = 8 × 7 × 6 × 5 × 4 × 3 × 2 × 1.) Perhaps we should really consider 6 = 3! to be superabundant as well. Note also that both 720 and 720720 are superabundant. The fact that 7 × 11 × 13 is 1001 will be discussed in the lecture on Notations.

n	prime factors	$s(n)$	$s(n)/n$	notes
12	$2^2.3$	16	1.333...	
24	$2^3.3$	36	1.5	4 factorial
36	$2^2.3^2$	55	1.527...	
48	$2^4.3$	76	1.583...	
60	$2^2.3.5$	108	1.8	first with 5
120	$2^3.3.5$	240	2.0	5 factorial
180	$2^2.3^2.5$	366	2.033...	
240	$2^4.3.5$	504	2.1	
360	$2^3.3^2.5$	810	2.25	
720	$2^4.3^2.5$	1698	2.358...	6 factorial
840	$2^3.3.5.7$	2040	2.428...	first with 7
1260	$2^2.3^2.5.7$	3108	2.466...	
1680	$2^4.3.5.7$	4272	2.542...	
2520	$2^3.3^2.5.7$	6840	2.714...	
5040	$2^4.3^2.5.7$	14304	2.838...	7 factorial
10080	$2^5.3^2.5.7$	29232	2.9	
15120	$2^4.3^3.5.7$	44400	2.936...	
25200	$2^4.3^2.5^2.7$	74744	2.966...	
27720	$2^3.3^2.5.7.11$	84600	3.051...	first with 11
...				
720720	$2^4.3^2.5.7.11.13$		3.509...	first with 13
...				
36756720	$2^4.3^3.5.7.11.13.17$		3.896...	first with 17

Some superabundant numbers.

n	primes	factors	$s(n)$	$s(n)/n$	notes
2	2	1	1	0.5	
3	3	1	1	0.333...	
4	2^2	1, 2	3	0.75	first composite
5	5	1	1	0.2	
6	2.3	1, 2, 3	6	1.0	perfect
7	7	1	1	0.142...	
8	2^3	1, 2, 4	7	0.875	
9	3^2	1, 3	4	0.444...	
10	2.5	1, 2, 5	8	0.8	
11	11	1	1	0.090...	
12	$2^2.3$	1, 2, 3, 4, 6	16	1.333...	first abundant
13	13	1	1	0.076	
14	2.7	1, 2, 7	10	0.714...	
15	3.5	1, 3, 5	9	0.6	
16	2^4	1, 2, 4, 8	15	0.937...	
17	17	1	1	0.058...	
18	2.3^2	1, 2, 3, 6, 9	21	1.166...	abundant
19	19	1	1	0.052...	
20	$2^2.5$	1, 2, 4, 5, 10	22	1.1	abundant
21	3.7	1, 3, 7	10	0.476...	
22	2.11	1, 2, 11	14	0.636...	
23	23	1	1	0.043...	
24	$2^3.3$	1, 2, 3, 4, 6, 8, 12	36	1.5	superabundant
25	5^2	1, 5	6	0.24	
26	2.13	1, 2, 13	16	0.615...	
27	3^3	1, 3, 9	13	0.481...	
28	$2^2.7$	1, 2, 4, 7, 14	28	1.0	perfect
29	29	1	1	0.034...	
30	2.3.5	1, 2, 3, 5, 6, 10, 15	42	1.4	abundant
32	2^5	1, 2, 4, 8, 16	31	0.968...	
36	$2^2.3^2$	1, 2, 3, 4, 6, 9, 12, 18	55	1.527...	superabundant
40	$2^3.5$	1, 2, 4, 5, 8, 10, 20	50	1.25	abundant
45	$3^2.5$	1, 3, 5, 9, 15	33	0.733...	
48	$2^4.3$	1, 2, 3, 4, 6, 8, 12, 16, 24	76	1.583...	superabundant
50	2.5^2	1, 2, 5, 10, 25	43	0.86	
60	$2^2.3.5$	1, 2, 3, 4, 5, 6, 10, 12, 15, 20, 30	108	1.8	superabundant
72	$2^3.3^2$	1, 2, 3, 4, 6, 8, 9, 12, 18, 24, 36	123	1.708...	

Some numbers and their factors.

The number system we use is a place system with base 10. This means that a number written as 1234 has value

$$1\times1000 + 2\times100 + 3\times10 + 4$$

so the value of each digit depends upon its place in the number.

Choosing a base for a number system involves various considerations. It is convenient if the base has plenty of factors because this makes it easier to divide into a number of equal portions. It is a tragedy that we have 10 fingers and toes and as a consequence have used 10 as the normal base for counting. It does not divide by 3 and so we get involved with recurring decimal fractions or approximations rather more often than is desirable.

If a base is small (such as binary or base 2) then it becomes cumbersome for even moderate numbers. Thus 27 in binary is 11011. Not only are such numbers hard to recognise but they take a lot of space to write. On the other hand if a base is large then we need lots of distinct symbols for individual digits. In base 16, we add the letters A to F to the normal decimal digits to represent the numbers 10 to 15. Thus 27 is 1B in hexadecimal notation. If a base is large such as 60 then we need 60 different characters or have to resort to compound notations using decimal numbers (for example) for the individual base 60 digits. Thus we can write a time as 12:20:17 meaning 12 hours, 20 minutes, 17 seconds. And we can write an angle as 12°20'17" meaning 12 degrees, 20 minutes, 17 seconds.

Probably the best base is obtained by choosing a small but abundant number. The obvious choice is 12 and so it is no coincidence that 12 has been used widely as a base. There are 12 inches to a foot and 12 points to a pica (the latter measurements are used in typesetting). 12 can be divided by 2, 3, 4, and 6. It is presumably also no coincidence that we have a special word for 12, namely a dozen. Goods are frequently sold in dozens. Eggs are a good example. Well actually they are usually sold in packs of 6 because they pack neatly two by three. When decimalization was in vogue, eggs were sold in tens in long boxes arranged as two by five. This just seemed to emphasise that 10 was a silly base. One supermarket now packs them two by two and three by three as well. We also have a special word for a dozen dozen, namely a gross = 144. Screws used to be sold by the gross but sadly now typically in boxes of 200. A great gross was a dozen gross, that is 1278.

If we wish to go beyond 12 and also divide by 5 then we find that 60 is the smallest base to do so. But 60 also divides by 6 and so we get a double bonus. It is not surprising therefore that 60 has proved to be a useful base and continues to do so. 60 was used by the Babylonians and, as mentioned above, is still used for time and angles – which are of course closely related.

We also have a special word for 20, a score. It is not used much these days except in phrases such as three score years and ten. The French also show a special treatment of 20 in that their word for 80 is quatre-vingt and for 90 is quatre-vingt dix – literally four score and ten. The Swiss say huitante and nonante. We will return to number bases in the lecture on Notations.

Weights and measures

IT IS INTERESTING to see what multiples are used or have been used in various measurements. Of course (almost) everything is going metric which seems easier although boring. However, we are the worse for it in some ways since, as we have seen, 10 has few factors and so is not an ideal base.

Here is a list of various items in ascending order.

2	2 pints = 1 quart	volume
	2 gallons = 1 peck	dry goods
	2 stones = 1 quarter	weight
	2 firkins = 1 kilderkin	beer
	2 kilderkins = 1 barrel	beer
3	3 kilderkins = 1 hogshead	beer
	3 barrels = 1 butt	beer
	3 feet = 1 yard	
	3 miles = 1 league	
	3 barleycorns = 1 inch	
	3 inches = 1 palm	
	3 scruples = 1 drachm	apothecaries weight
	3 quarters = 1 Flemish ell	cloth
4	4 gills = 1 pint	
	4 quarts = 1 gallon	
	4 inches = 1 hand	horses
	4 poles = 1 chain	distance
	4 roods = 1 acre	area
	4 quarters = 1 cwt	cwt = hundredweight, c = 100
	4 nails = 1 quarter	cloth (1 nail = 2¼ inches)
	4 quarters = 1 yard	cloth
	4 pecks = 1 bushel	dry goods
	4 bushels = 1 coomb	dry goods
5	5 quarters = 1 wey or load	dry goods
	5 quarters = 1 English ell	cloth
6	6 feet = 1 fathom	depth of water
	6 quarters = 1 French ell	cloth
	6 picas = 1 inch	typesetting
7	7 days = 1 week	
8	8 furlongs = 1 mile	
	8 drachms = 1 ounce	apothecaries weight
9	9 inches = 1 span	
	9 inches = 1 quarter	cloth
	9 gallons = 1 firkin	beer
10	10 chains = 1 furlong	distance

12	12 inches = 1 foot	
	12 points = 1 pica	typesetting
	12 ounces = 1 pound	apothecaries and troy
	12 semitones = 1 octave	music
14	14 pounds = 1 stone	
16	16 drams = 1 ounce	avoirdupois
	16 ounces = 1 pound	avoirdupois
	16 fl ounces = 1 pint	US
	16 sq poles = 1 sq chain	
20	20 grains = 1 scruple	apothecaries
	20 fl ounces = 1 pint	imperial
	20 cwt = 1 ton	
	20 dwt = 1 ounce	dwt = pennyweight, d = denarius,
	20 quires = 1 ream	paper
22	22 yards = 1 chain	
24	24 grains = 1 dwt	troy
	24 sheets = 1 quire	paper
	24 hours = 1 day	
28	28 pounds = 1 quarter	
36	36 bushels = 1 chaldron	dry goods
40	40 rods = 1 rood	square rods for area
60	60 minims = 1 fluid drachm	apothecaries
	60 seconds = 1 minute	
	60 minutes = 1 hour	
	60 minutes = 1 degree	
100	100 links = 1 chain	distance
112	112 lbs = 1 cwt	hundredweight is not a hundred!
640	640 acres = 1 sq mile	
1760	1760 yards = 1 mile	
4840	4840 sq yards = 1 acre	

Note that binary measures (2, 4, 8, 16) are very common indeed. This is particularly obvious with volumes. We have

4 gills = 1 pint
2 pints = 1 quart
4 quarts = 1 gallon
2 gallons = 1 peck
4 pecks = 1 bushel

There was also a pottle equal to two quarts and a chopin equal to two gills although these are now really archaic. Moreover, the term quartern was once used to mean a quarter of a pint. Actually it could mean a quarter of anything and we see it abbreviated as a quart to mean a quarter of a gallon.

The historic weights and measures which still linger in much of the English speaking world have a long history. A major definition was given by the Weights and Measures Act of 17 June 1824 during the reign of George IV – the official reference is CAP 74. Its title was "An act for ascertaining and establishing Uniformity of Weights and Measures". It covered length, weight, and volume.

There are two weight scales, one known as avoirdupois (from the old French for weight of goods) which is used for most things and another known as troy for precious metals and gemstones.

Curiously enough, it is the troy scale that was considered more fundamental. It is not binary but has 20 pennyweights to the ounce and 12 ounces to the pound thus exhibiting both a score and a dozen thus

24 grains = 1 dwt
20 dwt = 1 ounce
12 ounces = 1 pound

The term pennyweight probably arose from the fact that in the middle ages, a silver penny weighed 1/20 of an ounce. By contrast the George III cartwheel penny of 1797 weighed a whole ounce! The d in the abbreviation dwt for pennyweight comes of course from the d in denarius denoting the old sterling penny before decimalization in 1971.

The 1824 act defines the troy pound as that established by the physical standard made in 1758. Note the introduction of the grain; there are $24 \times 20 \times 12 = 5760$ grains to the pound troy. The use of 12 (a dozen) and 20 (a score) means that a troy pound can be amusingly described as two score gross grains!

The avoirdupois pound was then defined as precisely 7000 grains. The avoirdupois scale shows a strong binary influence. It is

16 drams = 1 ounce
16 ounces = 1 pound
14 pounds = 1 stone
2 stones = 1 quarter
4 quarters = 1 cwt
20 cwt = 1 ton

The 1824 act only defined the dram, ounce, and pound. The others seem to have come down by historic practice. It is not clear why there are 14 pounds to a stone. Indeed the number of pounds in a stone varied according to the item being weighed; we will come back to this later.

Beware that the troy ounce and avoirdupois ounce are not the same. The troy ounce is 480 grains whereas the avoirdupois ounce is 437.5 grains! So the troy ounce is more than the avoirdupois ounce but the troy pound is less than the avoirdupois pound. Madness!

Another scale is the related amusing old apothecaries' measure which also used the troy ounce

    20 grains = 1 scruple
    3 scruples = 1 drachm
    8 drachms = 1 ounce
    12 ounces = 1 pound

Note that there are 60 grains to a drachm. The corresponding liquid measures are almost identical with 60 minims to 1 fluid drachm and 8 fluid drachms to a fluid ounce. There are interesting old symbols for these measures thus

    Ə scruple   ʒ drachm   ℥ ounce

Glass measures for dispensing doses of (nasty dark) medicines were engraved with these symbols as shown below. The small glass on the left holds 120 minims (the engraving shows m for minim). The large glass holds 16 drachms or 2 fluid ounces. Both scales are just visible, that on the right has f (for fluid) followed by the drachm symbol, that on the left has the ounce symbol. Note also the modern plastic bottle with the historic symbol for ounce followed by the Roman number IV. The edge facing is marked in fluid ounces (up to 4); the other edge is in centilitres (to 10).

Some apothecaries' glass measures and a modern plastic bottle.

An important measure addressed by the 1824 act was the gallon. A major purpose of this act was to bring uniformity to some measures which had become rather uncertain and in particular to make the gallon such that a gallon of water weighed 10 pounds.

This is the origin of the difference between the US and Imperial gallon. In Britain we can remember the weight of a gallon because from childhood we will have learnt the rhyme

> A pint of water
> Weighs a pound and a quarter.

because there are 20 ounces to a pint. However, in the US they used to say

> A pint's a pound
> The world around.

Prior to 1824 the pint had been 16 ounces in many places including the United States. However, the US, no longer being a colony, stayed with the old size whereas the British Empire adopted the new unit. So the Canadian pint is the Imperial pint which is also used in Australia and New Zealand. Incidentally, I am told that disreputable bars in Canada might attempt to serve a US pint of beer. Anyway, it seems that the small old pint is not a pound the world around anymore!

A notable consequence is that US petrol is cheap by the gallon partly because it really is cheap but also because the gallon is smaller. Beware also if you buy a quart of gin in the US because it will be only 32 fluid ounces.

The above description has been slightly simplified. A fluid ounce is not quite the volume of an ounce of water and also there is a slight difference between the US fluid ounce (29.6 ml) and the UK fluid ounce (28.4 ml).

The Act of 1824 also defined the quart, pint, peck, and bushel. These measures were for liquids and certain dry goods. Normally, a measure full of stuff would have a flat top surface. In the case of liquids, gravity ensures this but otherwise the goods are stricken across to make the top level.

However, heaped measures were used for coal, lime, fish, potatoes, and fruit. The standard measure was a cylinder 19.5 inches in diameter and for heaped goods the heap had to be a cone at least 6 inches high. It must have been a source of many disputes.

It is notable that old decimal measures are rather rare. The only historic decimal measure is the chain measure used for surveying with 100 links to the chain and 10 chains to the furlong. It was devised in 1620 by the English clergyman Edmund Gunter (1581–1626). It was an important measure for military use overseas. One sidewalk in the town of Cambridge in New Zealand has two marks in the pavement giving the length of a chain.

Related to this are the curious rod, pole, and perch which are a quarter of a chain. The rod, pole, and perch are exactly the same but the different terms were used in different regions; the term perch seems to be completely obsolete but

A symbolic thin field one furlong long and one chain wide has an area of one acre. It is divided here into 4 roods of 40 square rods.

both pole and rod survive. The Act defined the yard as the fundamental length and then mentioned subdivisions into feet and inches and multiples giving furlongs and miles as well as the link, chain, and rod, pole, or perch.

A square rod is of course the area of a square plot whose sides are a rod. Extra confusion arises because square is sometimes omitted. A chain is 22 yards and so a link is exactly 7.92 inches. Chains are still used on some railways for measuring distances and especially the radius of curves. And the length of a cricket pitch is exactly a chain.

Note that a thin field of length one furlong and width one chain has an area of exactly one acre. And a square chain is one-tenth of an acre.

Small areas of land, such as used for allotment gardens, are still often measured in square rods and roods. There are 40 square rods to a rood and 4 roods to an acre. A full size allotment plot is typically 10 square rods. So for lengths we have

    100 links = 1 chain
    25 links = 1 pole
    4 poles = 1 chain
    10 chains = 1 furlong
    8 furlongs = 1 mile

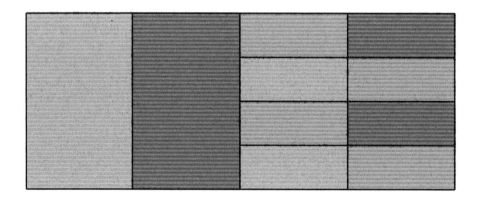

A symbolic rood of land divided into various allotments. The full plots are each 10 square rods: 4 rods by 2½. The quarter plots are 2½ square rods: 1 rod by 2½.

And for areas we have

> 40 sq rods = 1 rood
> 4 roods = 1 acre
> 160 sq rods = 1 acre
> 640 acres = 1 sq mile

In general, the term pole is used for lengths and rod (as an abbreviation for square rod) is used for areas.

Tape measures for surveying would typically be marked in inches and feet on one side and links and poles on the other. An ancient (and somewhat discoloured) tape measure is shown below. On the left is the tape inside its reel; the total length is given as 50 feet on the brass and leather case. The link side of the tape goes up to 75 links which is marked as 3 poles; the section below shows the part including 1 pole and 26 links.

Other curious measures are those for cloth and paper. The old cloth measure was binary in that it was based on multiples of 4 as follows

> 4 nails = 1 quarter
> 4 quarters = 1 yard

However, we note a degree of uncertainty regarding the ell which was 3 quarters if Flemish, 5 quarters if English, and 6 quarters if French. An old guide says "Scotch and Irish Linens, Woollens, Wrought Silks, Muslins, Cloths, Ribands, Cords, Tapes etc., are measured by the yard; Dutch Linens are bought by the Flemish ell but sold by the English ell; and Tapestry is sold by the Flemish ell." Clearly this provided opportunities for shady practices. An interesting fact according to Wightman's Arithmetical Tables is that the ell (from ulna, Latin for arm) was originally the length of the arm of King Henry I (1068–1135).

Traditional paper measure used both 20 and 24 thus

> 24 sheets = 1 quire
> 20 quires = 1 ream

so that a ream was 480 sheets. The term quire is still used but it now means 25 sheets so that a ream now consists of 500 sheets.

An ancient tape measure marked in feet and inches and also in links and poles.

Horses used to be measured in hands where a hand was four inches. These days they are still measured in hands but the hand is now defined as 10 cms.

The weights and measures Act of 1824 has now been replaced by definitions of the imperial units as equal to exact metric equivalents. Thus an inch is defined as exactly 2.54 cms so that an old hand was 10.16 cms.

There were many other obscure measures. Herrings for example could be measured by number

   4 herrings = 1 warp
   66 warps = 1 cut
   5 cuts = 1 last

So a last was 1320 herrings. Herrings as caught were measured by the volume known as a cran which was a barrel of some 40 gallons (but which gallon one might ask). A cran would be 1000 to 2000 (uncleaned) herrings according to their size. Maybe a last was roughly a cran of herrings after they had been cleaned.

Another curious set of measures was for the weight of wool thus

   7 pounds = 1 clove
   2 cloves = 1 stone
   2 stones = 1 tod
   6½ tods = 1 wey
   2 weys = 1 sack
   12 sacks = 1 last

So there are 14 pounds to a stone and 26 stones to a sack. This definition of a sack and a stone was by an act of Edward III. Note again the definition of a last for a large quantity of wool – not to be confused with a last of herrings.

Although a stone of wool was 14 pounds, a stone of other goods was often different. A stone of cheese was 16 pounds, a stone of glass was 5 pounds, and a stone of meat was 8 pounds.

In a similar way the term last was clearly used to mean a large amount – again according to context. For dry goods the measures go on thus

   4 bushels = 1 coomb
   2 coombs = 1 quarter
   5 quarters = 1 wey or load
   2 weys = 1 last

We noted earlier that binary measures are common and one wonders why. Perhaps because we naturally divide things into two. Dividing a lump of cake into two is easy; one person cuts and the other chooses. But there are no similar easy ways of cutting a cake into three portions.

Thus we might ask for half a pound of sausages, a quarter of a pound of cheese, and so on (of course the butcher has to measure the stuff in kilos). We are

Section of a foot rule showing binary subdivisions.

less likely to be so precise as to ask for 6 ounces of something. And if we asked for a third of a pound of bacon, the butcher would look very cross.

Binary subdivisions show up on foot rules as well. Inches are typically divided into 16ths with different length marks according to the subdivision. Thus a long mark for a half, a lesser mark for the quarters and so on. The first inch might be divided into 32nds with even tinier marks as in the diagram above.

Using such binary measures can be tedious. We might have drills of sizes $^1/_2$, $^7/_{16}$, $^{15}/_{32}$, and $^{31}/_{64}$ inch. It is not immediately obvious how they are ordered – if $^{15}/_{32}$ is not quite big enough, which should I try? However, metric drills of sizes 10, 11, 12, and 13 mm are obviously in ascending order.

In the UK, about the only official uses for the old measures are pints for beer and milk, and miles for distances. But the old measures leave their shadow. Timber is sold in metres but only in lengths close to multiples of a foot such as 2.1, 2.4, and 2.7 metres. One cannot buy a plank 2 metres long! And the widths of paint brushes are still given in inches.

If you feel passionately that we should count in twelves then join the Dozenal Society. Incidentally, the metric system was devised soon after the French revolution by a committee chaired by the French mathematician Joseph-Louis Lagrange (1736–1813) and included Pierre-Simon de Laplace (1749–1827) as a member.

The committee also defined temporary "mesures usuelles" to help with the transition in French as follows.

une toise	2 metres, about a fathom for depths,
une aune	120 cms, about an ell for cloth,
un boisseau	12.5 litres, about three gallons for volume,
une livre	500 grams, about a pound for weight.

It is a bit surprising that the aune was roughly an English ell rather than a French ell!

The livre survives in common use. Note that it is not a book which is masculine whereas the weight is feminine.

It is clear that metrication is here to stay for most things but there is one rather special measurement which is basically binary and will no doubt stay that way for ever. This is the length of musical notes and deserves a brief discussion.

Musical notes

THE BASIC LENGTHS of musical notes form a binary hierarchy consisting of crotchet, quaver, semiquaver, and so on but there are curious rules regarding intermediate lengths. The notes are as follow

    2 minims = 1 semibreve
    2 crotchets = 1 minim (1/2)
    2 quavers = 1 crotchet (1/4)
    2 semiquavers = 1 quaver (1/8)
    2 demisemiquavers = 1 semiquaver (1/16)
    2 hemidemisemiquavers = 1 demisemiquaver (1/32)

Once upon a time there was a breve being two semibreves but it is rarely encountered now. The US terminology reflects the subdivision in numerical terms. The semibreve is known as a whole note, the minim as a half note, and the crotchet as a quarter note and so on.

Music is divided into bars to define the rhythm and the first beat of each bar is emphasized. The rhythm is indicated by a time signature at the start of the piece which looks rather like a fraction where the numerator indicates the number of beats and the denominator indicates the length of each beat.

A frequent rhythm is that with 4 crotchets to a bar (written $\frac{4}{4}$) and so the length of the bar is a semibreve which as mentioned above is also known as a whole note. A bar of four crotchets thus might be written as shown in the first bar below. The second bar shows two minims and the third shows a semibreve.

If a note is followed by a dot then its length is extended by half. Thus a dotted minim has the length of three crotchets as shown in bar 4; the bar is completed by a crotchet. A second dot adds half the value of a previous dot so that a double dotted minim has the length of seven quavers as shown in bar 5; the bar is completed by a quaver.

Another way of extending a note is by using a tie to another note of the same pitch; the second note is not sounded. Thus if we wanted a note of length five quavers, then we can tie a minim to a quaver. The first bar on the second line shows a minim tied to a quaver followed by a dotted crotchet; if a dotted note is on a stave line then the dot is in the space above. Several notes could be tied together as shown next where the last crotchet in one bar is tied to the semibreve in the next bar and then to the first crotchet in the following bar.

Further variety is introduced by the range of subdivisions of a bar. The commonest time signature is $\frac{4}{4}$ which is a typical march or quickstep and so is often known as common time. Another frequent time signature is $\frac{3}{4}$ which is a waltz consisting of three crotchets to a bar.

Sometimes the style of music demands that a note be subdivided into three smaller notes. Thus we might wish for a crotchet to be subdivided into three equal notes. These are represented by quavers and the variation indicated using a slur with a 3 as shown in the first bar above.

An alternative approach is to recognise that there are six quavers to the bar and use $\frac{6}{8}$ rather than $\frac{2}{4}$ as shown in the second bar above; the crotchets are of course now dotted so that they equal three quavers. This is called a compound time.

There are many other possibilities involving powers of two and three, thus Chopin's Nocturnes also use $\frac{3}{4}$, $\frac{4}{4}$, $\frac{6}{8}$, $\frac{9}{8}$, and $\frac{12}{8}$. Ecclesiastical music often uses minims for individual beats and so we also find $\frac{3}{2}$, $\frac{3}{2}$, and $\frac{4}{2}$. The last is known as *Alla Breve*; the length of the bar is four minims which equals a breve.

Almost all music involves rhythms with subdivisions into two, three, or four. But sometimes subdivisions into five or seven are found. An example of subdivision into five is found in Mars from *The Planets* suite by Holst. However, it sounds a bit like alternate bars of three and two beats. Similarly bars of seven have the effect of alternate bars of four and three beats.

Well, that is quite enough about musical timing. But it clearly illustrates our strong preference for division by twos and threes. Metrication and division by ten is just not relevant.

Currencies

OTHER SOURCES of interesting measurements are some old currencies. Most currencies have been decimal for a long time and the only major ones that were not decimal until recently are those of Britain, India, and Cyprus. Thus in Britain we had

    4 farthings = 1 penny
    12 pence = 1 shilling
    20 shillings = 1 pound

A value such as 3 pounds, 4 shillings, and 5 pence was written as

£3-4s-5d.

The symbol £ is really an L from the Latin word libra meaning a pound (the value of a pound of silver perhaps a long time ago), the s is for solidus (originally a Roman gold coin, and then in medieval England a shilling), and the d is for denarius (originally another Roman coin).

It was a flexible currency with 240 pence to the pound. 240 is a superabundant number and the pound divided easily and precisely into convenient fractions. Thus if a salary is a multiple of a pound and a pension contribution is 5% per month then the deduction is exactly 1 penny per month per pound of salary. Hence if a salary is £10,000 per annum then the monthly deduction is precisely £41-13s-4d. But in decimal currency the deduction is £41.666666666666... and tiny errors can arise which can keep the auditors amused for ages. The key point was that the 12 pence in a shilling matched the fact that there are 12 months in a year. Of course, we also had nice things such as guineas and groats. It seems that horses are still sold in guineas. A guinea is 21 shillings (which in modern terms is £1.05). A groat was 4d so there were 60 groats to the pound.

The various British Dominions such as Australia and New Zealand which still used shillings and pence went decimal in the late 1960s and Britain and the remaining colonies finally went decimal in 1971.

Although Britain was rather slow in going decimal, it had been thinking about the topic for many years. Indeed, the florin (equal to 2 shillings) was introduced in 1849 as a first step towards decimalization there being 10 florins to a pound. A double florin was issued in 1887 but was unpopular.

An interesting discussion on the problems of decimalization will be found in *Arithmetic* by the mathematician John William Colenso (1814–1883). This was first published in 1843 but a revised version in 1874 and then 1886 contains interesting appendices on decimalization and the metric system. It is obvious to us that the correct approach was to define a new unit (the new penny) with 100 such pennies to the pound. However, an alternative approach was based on the farthing of which there were 960 to the pound. The thought was that the pound should be redefined to be 1000 farthings thus

10 farthings = 1 doit
10 doits = 1 florin
10 florins = 1 pound.

Although keeping the farthing would have been helpful to the poor, changing the pound would have been awkward for the wealthy. Inevitably, the wealthy won but it took about a century to happen.

Colenso had an interesting career, he was of a humble background but managed to get to St John's College, Cambridge to read mathematics. He was one of Mr Hopkins' men. William Hopkins (1793–1866) was a private tutor who crammed men for the Mathematical Tripos examination. (It is called the Tripos because at one time students sat on a three-legged stool for part of the ordeal.)

Hopkins tutored 19 Senior Wranglers between 1829 and 1851. (A wrangler is someone who obtained a first in mathematics at Cambridge; Senior Wrangler is the top one in the year.) Colenso was Second Wrangler in 1836. He was ordained and was a master at Harrow for some time. He later became Bishop of Natal and the small town of Colenso near Ladysmith was named after him. There was a major Battle of Colenso in 1899 during the Boer War. A facsimile of the 1886 version of his book on arithmetic was recently republished by Bibliolife.

Another interesting currency was that of India which was

12 pies = 1 anna
16 annas = 1 rupee

and that also had nice divisibility properties. But note that Ceylon (Sri Lanka) had 100 cents to 1 rupee as long ago as 1872; also Mauritius. Several other countries also used annas and rupees such as Iraq, Somalia, Somaliland, Zanzibar, Kenya, Uganda and Tanganyika. An oddity in India was Travancore with 16 cash to 1 chuckram and 28 chuckrams to 1 rupee.

India went decimal in 1957 with 100 paise to 1 rupee. The East African countries (Kenya etc.) went decimal in 1922 with 100 cents to 1 shilling (and then 20 shillings to 1 pound although the pound died in about 1960) – so the shilling lives on.

There were some other strange currencies in the middle of the nineteenth century. For example the German region had currencies such as 60 kreuzer to 1 gulden in Bavaria and Prussia, 60 kreuzer to 1 florin in Austria, 60 silbergroschen to 1 thaler in Bremen (but only 30 silbergroschen to 1 thaler in Brunswick, Mecklenburg-Strelitz, Oldenburg, and Hanover), 16 schillings to 1 mark in Hamburg, Schleswig-Holstein, and Heligoland (but only 15 in Lubeck). One wonders whether the 60 has any connection with the use of 60 in Babylonia.

There were 40 paras to 1 piastre in Turkey, Syria, Saudia Arabia, Egypt, Roumania, Albania, and Cyprus. Cyprus was unusual in also having 180 piastres to £1 which continued until 1955 when it changed to 1000 milliemes to £1 and then in 1983 to 100 cents to £1. Note also the Yemen had 40 bogaches to 1 imadi! Egypt also had 1000 milliemes to £1 but now has 100 piastres to £1.

In Scandinavia there were 96 rigsbank skilling to 1 rigsdaler in Denmark and Iceland, but 120 skilling to 1 rigsdaler in Norway and 48 skilling to 1 rixdaler in Sweden.

The historic Spanish coinage is interesting and was the foundation of the currencies in the Spanish colonies in America. It was based on the reale (from royal). The eight reale coin was often cut into eight pieces from which we get the term pieces of eight. The doubloon (from the Spanish for double) was a gold coin equal to 32 reales. Until recently many countries in Central and Southern America had 8 reales to 1 peso. Examples were Costa Rica, Dominican Republic, Ecuador, Guatemala, Honduras, Mexico, Venezuela, Paraguay, and Salvador.

A Maria Theresa thaler; note the standard date of 1780.

A 1959 Guernsey penny inscribed eight doubles.

It should also be noted that many currencies and names are derived from the silver Maria Theresa thaler which circulated widely around the world. Thus dollar, rixdaler, and so on are all variants of the word thaler. Although the Empress of Austria died in 1780, coins continued to be struck inscribed 1780 for many years; indeed some were even struck in the Austrian mint as recently as 2003. In some areas the silver thaler was recognized as the only safe money to have when travelling. Poor camels crossed seemingly endless deserts laden with baskets and sacks of thalers. An excellent quality thaler is shown above.

Before decimalization, the currency in the Channel Island of Guernsey was also based on pounds, shillings, and pence with the curious variation that the penny was subdivided into eight doubles. Indeed, the Guernsey penny of 1959 is inscribed eight doubles as shown above. Note that this double was nothing to do with the doubloon but related to being twice an old French coin.

Sadly, all these amusing divisions have been swept away with global decimalization so that probably the whole world now uses 100 This to 1 That.

Further reading

A THOROUGH ACCOUNT of Euclid's Elements is *The Thirteen Books of Euclid's Elements* by Sir Thomas L Heath. This is in three volumes and includes much commentary – originally published in 1908, it is now published by Dover. The proof that prime numbers go on for ever is in Volume II.

For the weird formula that generates primes and related matters see the discussion at https://en.wikipedia.org/wiki/Formula_for_primes.

A convenient source of information on old currencies is *Gibbons Simplified Stamp Catalogue*.

Mr Hopkins and his wranglers are described in *Mr Hopkins' Men* by Alex Craik. For arithmetic and the problems of decimalization see *Arithmetic Designed for the Use of Schools: To which is Added a Chapter on Decimal Coinage* by John William Colenso; a facsimile is published by Bibliolife.

For information about the town of Colenso which actually has its own flag see http://en.wikipedia.org/wiki/Colenso,KwaZulu-Natal.

For weights and measures and especially the apothecaries' madness see http://en.wikipedia.org/wiki/Apothecaries'_system. Also, the little booklet *Wightman's Arithmetical Tables* is fascinating; it was originally written in about 1840 but a revised version by Frank Sandon was certainly published in 1941.

For a fascinating description of the origins and use of the Maria Theresa thaler see *A Silver Legend, the Story of the Maria Theresa Thaler* by Clara Semple.

See also Appendix A for a discussion of Ackermann's function.

Exercises

1. Find the factors of 220. Show that they add up to 284. Then find the factors of 284 and add them up. Comment on the result.

2. The numbers 101, 103, 107, 109 are all prime. Find the next such decade with four primes. Show that such decades of four primes cannot be adjacent.

3. A lorry in 1960 has its unladen weight marked as 2 tons 5 cwt 2 qtrs 10 lbs. A firm has a fleet of three such lorries. They have loads of 5 cwt 2 qtrs 22 lbs, 4 cwt 3 qtrs 0 lb and 6 cwt 0 qtr 25 lbs respectively. What is the gross weight of each lorry and what is the sum of all their weights?

    They pass over a toll bridge where the toll is 1/– per cwt for the unladen weight and 1/3 per qtr for the carried goods. (Fractions of cwt and qtr are rounded up as appropriate.) How much is the toll (including the load) per lorry and how much is the total toll?

    Note that 1/– means one shilling and 1/3 means one shilling plus three pence.

4*. Show that $S_x \equiv 1 + 1/x + 1/x^2 + 1/x^3 + 1/x^4 + \cdots = x/(x-1)$. Hence $S_2 = 2$, $S_3 = 3/2$, $S_5 = 5/4$ and so on.

    Deduce the maximum value that $s(n)/n$ can have for a number whose only factors are powers of 3 and 5.

    Conclude that such a number cannot be abundant. Note that the smallest odd abundant number (945) has factors of 3, 5, and 7.

    Is it possible for an odd abundant number not to have a factor of 3? Justify your answer. Find an odd abundant number without a factor of 5.

2 Amicable Numbers

IN THIS LECTURE we look a little more at perfect numbers and the closely related amicable and sociable numbers.

Perfect numbers were known to Euclid and were later studied by Mersenne the Monk whose name is associated with certain prime numbers and who also explored musical scales as we shall see in a later lecture.

Under the banner of amicable in the sense of Friendly, we also say a little about Fermat numbers and Fibonacci numbers; the latter in particular will crop up several times in later lectures.

Perfect numbers

WE RECALL that a number whose factors (apart from the number itself) add up to the number is called a perfect number.

The first two perfect numbers are 6 and 28 since

$1 + 2 + 3 = 6$

$1 + 2 + 4 + 7 + 14 = 28$

Some mystical properties were attached to these perfect numbers – God made the world in 6 days and the Moon encircles the Earth in 28 days.

Two more were known in antiquity, namely

$1 + 2 + 4 + 8 + 16 + 31 + 62 + 124 + 248 = 496$

$1 + 2 + 4 + 8 + 16 + 32 + 64 +$
$\quad 127 + 254 + 508 + 1016 + 2032 + 4064 = 8128$

The next perfect number was discovered in the 15th century but the author seems to be unknown. It is quite a big jump, namely

33,550,336

The factors of these perfect numbers are interesting, they are

$6 = 2 \times 3$
$28 = 2^2 \times 7$
$496 = 2^4 \times 31$
$8128 = 2^6 \times 127$
$33,550,336 = 2^{12} \times 8191$

We see a pattern emerging. The factors of these perfect numbers all comprise lots of powers of two and one other prime number. Moreover, this other prime number is itself one less than a power of 2. So we have

$6 = 2^1 \times (2^2-1)$
$28 = 2^2 \times (2^3-1)$
$496 = 2^4 \times (2^5-1)$
$8128 = 2^6 \times (2^7-1)$
$33{,}550{,}336 = 2^{12} \times (2^{13}-1)$

So these perfect numbers all have the form $2^{k-1} \times (2^k-1)$. But note that not every number of this form is perfect. It never works if k is even. For example taking $k = 4$ we get $8 \times 15 = 120$ and this is certainly not perfect.

Euclid showed that if 2^k-1 is a prime number then $2^{k-1} \times (2^k-1)$ is always a perfect number. His proof is Proposition 36 in Book IX of the Elements. But as in the case of showing that primes go on for ever, his discussion is geometrical and very difficult to follow.

Before describing a modern proof it is useful to introduce some algebraic notation. We will denote the sum of all the factors of n by $\sigma(n)$. This includes n itself. Note therefore that $\sigma(n) = s(n) + n$ where $s(n)$ is the sum of all the factors except n itself as introduced in the previous lecture.

So a number is perfect if

$\sigma(n) = 2n$ rule for n to be perfect

A very useful fact is that if two numbers m and n have no common factors (we say they are relatively prime or coprime) then

$\sigma(m \times n) = \sigma(m) \times \sigma(n)$ provided m and n have no common factors

Functions having this property are known as multiplicative functions. We will meet another example in the final lecture.

We can try this multiplicative property on an example. Consider 3, 4, and 12. Now 3 and 4 have no common factors and

$\sigma(3) = 1 + 3 = 4$
$\sigma(4) = 1 + 2 + 4 = 7$
$\sigma(12) = 1 + 2 + 3 + 4 + 6 + 12 = 28 = 4 \times 7$

So it works. It is not difficult to see why this is the case. Consider writing

$(1 + 3) \times (1 + 2 + 4)$

then every factor of 12 is obtained by taking one item from the first term and one item from the second term. Thus 6 is obtained by taking the 3 from the first term and the 2 from the second. But it only works if the two numbers have no common factors because otherwise we get duplication.

Now to return to Euclid's proposition. We have

$$\sigma(2^{k-1}) = 1 + 2 + 4 + \cdots + 2^{k-1} = 2^k - 1$$
$$\sigma(2^k - 1) = 1 + 2^k - 1 = 2^k$$

The first is because we are just adding up the powers of 2 such as $1 + 2 + 4 = 7$. The second is because we are assuming that $2^k - 1$ is prime and so its only factors are 1 and itself. Moreover, since $2^k - 1$ is prime it has no factors in common with 2^{k-1} and so we can apply the multiplication rule that $\sigma(m \times n) = \sigma(m) \times \sigma(n)$ to these two expressions. We get

$$\sigma(2^{k-1} \times (2^k - 1)) = \sigma(2^{k-1}) \times \sigma(2^k - 1) = (2^k - 1) \times 2^k$$

Now the expression on the right is simply twice that following σ on the left which is precisely the condition for being perfect. So we are done.

A further important fact is that not only are all numbers of Euclid's form perfect but all even perfect numbers are of that form which is perhaps surprising. This was shown by Euler about 2000 years later. The somewhat tricky proof is roughly as follows.

First we extract all the powers of two so that the perfect number n has the form $2^{k-1} \times m$ where m is odd. As a consequence m has no factors in common with 2^{k-1} and so the multiplication rule can be applied. We get

$$\sigma(n) = \sigma(2^{k-1}) \times \sigma(m) = (2^k - 1) \times \sigma(m)$$

On the other hand since we are given that n is a perfect number we know that $\sigma(n) = 2n = 2^k \times m$. Equating these two expressions for $\sigma(n)$ we get

$$(2^k - 1) \times \sigma(m) = 2^k \times m$$

It follows that $2^k - 1$ must be a factor of m since it has to be a factor of the right hand side and it is obviously not a factor of 2^k (it is odd for one thing). So m must be of the form

$$m = (2^k - 1) \times M$$

Putting this expression for m in the previous equation and cancelling the factor of $2^k - 1$ gives

$$\sigma(m) = 2^k \times M = 2^k \times M - M + M = (2^k - 1)M + M = m + M$$

Now we know that M is a factor of m and moreover m is a factor of m. Therefore $\sigma(m)$ must be at least $m+M$. But since $\sigma(m)$ is exactly $m+M$ there can be no other factors. However, 1 must be a factor and so we conclude that M must be 1 and that m must be prime. That's it. Indeed, it is a somewhat crafty proof!

We have just shown that all even perfect numbers must be of the form

$$n = 2^{k-1} \times (2^k - 1) \text{ where } 2^k - 1 \text{ is prime}$$

What about odd perfect numbers? No odd perfect numbers are known but it has not been proved that there are none. But it seems unlikely. If there are odd perfect numbers then they must be extremely large.

In the search for perfect numbers it was thought for some time that if p were prime then 2^p-1 would be prime as well. It certainly works to start with and gives the first four perfect numbers. Thus

$$2^2-1 = 3$$
$$2^3-1 = 7$$
$$2^5-1 = 31$$
$$2^7-1 = 127$$

and these are all prime. But it is not true for 11 since

$$2^{11}-1 = 2047 = 23 \times 89$$

However, $2^{13}-1 = 8191$ is prime and so the fifth perfect number is $2^{12} \times (2^{13}-1)$ which is 33,550,336 as mentioned earlier.

Modular arithmetic

BEFORE DIVING into perfect numbers in more detail it is helpful to introduce the idea of modular arithmetic. A good example occurs with using a clock (in 12 hour notation).

If the time is 8 o'clock, then what is the time 7 hours later? It is 3 o'clock. And if the time now is 11 o'clock then what is the time 15 hours later? It is 2 o'clock. We throw away multiples of 12 whenever they occur. The numbers just go around in cycles.

We can do this modulo any integer greater than 1. For example we could be working modulo 8 in which case if we add 7 to 5 we get 4. That is because $7 + 5$ gives 12 and then we subtract 8 to give 4. We can write this as

$$7 + 5 \equiv 4 \pmod{8}$$

We do the same with other operations such as multiplication. Thus since 7 times 5 equals 35 and $35 = 4 \times 8 + 3$, we have

$$7 \times 5 \equiv 3 \pmod{8}$$

We read such statements as "7 times 5 is congruent to 3 mod 8". Note the special symbol $\equiv$ rather than the normal $=$ which is used to distinguish congruence from normal equality.

The key definition is

$a \equiv b \pmod{m}$ means $a-b$ is exactly divisible by m

It is easy to show various simple consequences, such as

$a \equiv b \pmod{m}$ implies $b \equiv a \pmod{m}$

$a \equiv b \pmod{m}$ and $b \equiv c \pmod{m}$ implies $a \equiv c \pmod{m}$

The congruency property is preserved by addition, subtraction, and multiplication. So if

$a \equiv b \pmod{m}$ and $c \equiv d \pmod{m}$

then it follows that

$a + c \equiv b + d \pmod{m}$

$a - c \equiv b - d \pmod{m}$

$a \times c \equiv b \times d \pmod{m}$

$a^n \equiv b^n \pmod{m}$

But it does not hold for division. For example 3 is congruent to 13 modulo 10 but dividing 12 by 3 and 13 gives different results thus

$12 \div 3 = 4$ rem 0 but $12 \div 13 = 0$ rem 12

and neither quotient nor remainder are congruent modulo 10.

However, if $a \equiv b \pmod{m}$ and d is a divisor of m then it follows that $a \equiv b \pmod{d}$. For example, since $25 \equiv 15 \bmod 10$, it follows that $25 \equiv 15 \bmod 5$ and also that $25 \equiv 15 \bmod 2$.

Moreover, if a and m are relatively prime (and so have no common factors) and $ab \equiv ac \pmod{m}$ then we can cancel the a and deduce that $b \equiv c \pmod{m}$.

Also, although division does not preserve congruency in general, a always has an inverse modulo m if a and m are relatively prime, that is we can always find b such that $a \times b \equiv 1 \pmod{m}$. For example, 3 and 7 have no common factors and $3 \times 5 \equiv 1 \pmod{7}$ so that 5 is the "inverse" of 3. This will be proved and used in the final lecture when we discuss cryptography.

Mersenne the monk

THE FRENCH MONK Father Merin Mersenne (1588–1648) was also a famous mathematician. Among many topics he studied numbers of the form

$M_n = 2^n - 1$

which are now called Mersenne numbers. If a Mersenne number is prime then it is called a Mersenne prime. Note that it can be proved that if a Mersenne number

M_n is prime then n must be a prime (see the end of this section). But the reverse does not hold since we saw above that M_{11} is not prime.

We have seen that all even perfect numbers are of the form

$$2^{k-1} \times (2^k-1) \text{ where } (2^k-1) \text{ is prime}$$

and all numbers of this form are perfect.

In other words all perfect numbers are associated with a Mersenne prime and vice versa. Mersenne asserted that M_p was prime for the following values of p: 2, 3, 5, 7, 13, 17, 19, 31, 67, 127, 257. Why he stated this is not known. It was only in 1947 that it was confirmed that he was wrong. The cases of $p = 61, 89,$ and 107 are also prime but 67 and 257 are not.

There is an interesting test for whether a Mersenne number is in fact a prime which was devised by Lucas and Lehmer. The theory was developed by the French teacher Edouard Lucas (1842–1891) and a simple test was devised by the American mathematician Derrick Lehmer (1905–1991). The test goes as follows. We form the series of numbers

$$L_{i+1} = (L_i)^2 - 2, \text{ starting with } L_2 = 4$$

and then M_p is prime if and only if L_p is exactly divisible by M_p. (Some texts define L_0 as 4 and then the test is that M_p is prime if and only if L_{p-2} is exactly divisible by M_p but the difference of 2 is confusing so we have omitted it.)

Let's try this on M_3 and M_5. The first few Ls are

$L_2 = 4$
$L_3 = L_2^2 - 2 = 16 - 2 = 14$
$L_4 = L_3^2 - 2 = 14^2 - 2 = 196 - 2 = 194$
$L_5 = L_4^2 - 2 = 194^2 - 2 = 37636 - 2 = 37634$

Now $M_3 = 2^3 - 1 = 7$. Is $L_3 = 14$ divisible by 7? Yes, so M_3 is prime. And from this we get 28 is perfect.

Similarly $M_5 = 2^5 - 1 = 31$. Is $L_5 = 37634$ divisible by 31? Yes it is (37634 = 31 × 1214) so M_5 is prime. And from this we get 496 is perfect.

The numbers rapidly get very large indeed

$L_6 = L_5^2 - 2 = 37634^2 - 2 = 1416317956 - 2 = 1,416,317,954$
$L_7 = L_6^2 - 2 = 1416317954^2 - 2 = 2,005,956,546,822,746,114$

Now $M_7 = 2^7 - 1 = 127$. Is L_7 divisible by 127? Yes it is since we have

$$L_7 = 2,005,956,546,822,746,114 = 127 \times 15,794,933,439,549,182$$

and so $M_7 = 127$ is prime. This gives 8128 is perfect.

The next Mersenne number to try is $M_{11} = 2047$. As we saw earlier this is 23×89 and so is not prime. And this is confirmed by the Lucas–Lehmer test. We have

2 Amicable Numbers   31

$L_{11} =$
 68,729,682,406,644,277,238,837,486,231,747,530,924,247,154,108,
 646,671,752,192,618,583,088,487,405,790,957,964,732,883,069,102,
 561,043,436,779,663,935,595,172,042,357,306,594,916,344,606,074,
 564,712,868,078,287,608,055,203,024,658,359,439,017,580,883,910,
 978,666,185,875,717,415,541,084,494,926,500,475,167,381,168,505,
 927,378,181,899,753,839,260,609,452,265,365,274,850,901,879,881,
 203,714

and this equals 2047 times

 33,575,809,675,937,604,904,170,730,938,811,690,729,969,298,538,
 664,715,071,906,506,391,347,575,674,543,701,985,702,434,327,846,
 878,868,313,033,543,691,057,729,380,731,463,895,904,418,469,015,
 419,986,745,519,437,033,734,832,938,279,608,910,120,948,160,191,
 000,813,964,765,860,974,861,301,658,488,764,277,072,487,136,544,
 175,563,352,173,792,789,086,765,731,443,754,408,818,222,706,341,
 574

plus a remainder of 1736. So since there is a remainder this confirms that $M_{11} = 2047$ is not prime.

The numbers are getting a bit large now. For $M_{13} = 8191$ we find that

$L_{13} =$
 22,313,995,867,897,900,769,603,796,342,295,788,566,208,710,409,
 165,129,831,038,160,968,311,491,946,220,319,893,407,703,857,721,
 437,957,260,314,978,754,160,034,503,401,040,789,215,400,628,158,
 170,099,668,522,698,066,550,221,265,307,171,574,634,992,724,727,
 060,201,120,758,890,920,172,789,110,609,085,990,337,846,018,634,
 451,646,739,004,908,975,710,893,057,017,831,784,106,285,989,578,
 600,398,251,364,366,079,398,506,512,806,386,775,247,318,462,388,
 007,386,288,293,644,987,819,640,076,171,556,974,003,404,195,908,
 596,970,825,853,990,347,990,259,288,695,088,334,854,125,701,652,
 040,860,084,239,663,064,263,605,520,384,355,127,215,307,437,936,
 591,866,962,296,906,419,378,104,850,899,571,605,034,504,288,737,
 636,564,836,267,334,726,723,727,575,106,663,971,046,844,142,763,
 512,854,023,937,849,655,467,693,015,631,287,929,701,909,077,381,
 005,060,802,853,209,341,459,156,871,829,180,256,316,747,660,704,
 875,518,660,035,573,112,882,904,493,746,617,877,304,844,878,674,
 402,542,586,943,400,547,464,667,179,926,000,026,596,616,252,849,
 884,072,241,228,637,895,801,783,293,732,168,802,374,542,280,341,
 992,348,946,606,531,635,000,814,995,246,895,089,041,641,203,184,
 136,132,975,956,905,572,518,723,976,402,989,858,509,003,359,081,
 748,048,869,560,319,466,898,146,867,908,972,088,453,016,102,089,
 761,833,396,052,479,183,215,782,590,173,494,080,725,569,259,056,

977,955,738,902,892,341,951,393,866,495,222,420,379,013,713,784,
627,095,469,233,910,359,313,068,881,745,808,900,306,832,764,925,
725,008,680,492,006,161,979,334,986,865,505,218,272,485,914,888,
669,136,966,553,469,714,434

and clearly this exactly equals 8191 times

2,724,208,993,761,189,203,956,024,458,832,351,186,205,434,062,
894,045,883,413,278,106,252,166,029,327,349,516,958,576,957,358,
251,490,326,005,979,581,755,589,610,963,379,415,116,029,865,481,
402,771,293,922,927,367,421,587,262,276,543,959,789,402,115,093,
036,283,862,868,867,161,539,835,076,377,620,069,629,818,827,815,
218,123,152,118,777,801,942,484,807,351,706,969,125,416,431,397,
704,846,569,572,013,927,407,948,542,645,145,498,137,873,087,826,
639,895,774,422,371,503,823,665,007,468,142,714,443,096,593,322,
988,276,257,581,979,043,827,403,160,626,918,365,871,581,699,627,
889,251,627,913,522,532,567,892,262,285,966,930,437,713,031,123,
988,751,918,239,153,512,315,725,167,976,995,678,798,010,534,579,
127,892,178,765,393,081,030,854,300,464,737,391,166,749,376,481,
932,957,395,182,254,871,867,622,148,166,437,300,659,493,233,717,
617,514,443,029,326,009,212,447,426,666,973,538,800,726,121,438,
759,067,105,363,883,910,741,411,853,710,977,643,426,302,634,437,
114,215,918,318,080,887,249,989,888,893,419,610,132,659,779,373,
688,691,520,110,931,253,302,622,792,544,520,669,316,877,338,584,
054,736,777,756,871,155,536,664,020,906,714,087,296,012,843,753,
404,484,553,284,935,364,731,867,168,404,711,251,191,430,027,967,
494,573,174,161,923,997,912,116,575,254,422,181,473,936,772,322,
031,721,816,146,072,418,900,718,177,288,913,939,778,484,832,017,
699,664,966,292,625,118,050,469,279,269,347,139,589,673,265,020,
708,960,501,676,707,405,605,306,907,794,629,337,114,739,685,621,
502,259,636,246,124,546,695,072,028,673,605,813,487,057,247,575,
225,141,858,937,061,374

and as a consequence we confirm that 8191 is prime and this gives rise to the fifth perfect number which as we have seen is 33,550,336.

Actually we don't have to work with such huge numbers since for example if we want to know whether L_5 is divisible by M_5 (= 31) we can keep taking the remainder on dividing by M_5 at each stage. In other words we work entirely using modular arithmetic with modulo 31 as explained in the previous section. The calculation for M_5 then goes

$L_2 = 4$
$L_3 = L_2^2 - 2 = 16 - 2 = 14$
$L_4 = L_3^2 - 2 = 196 - 2 = 194 \equiv 8 \bmod 31$
$L_5 = L_4^2 - 2 = 64 - 2 = 62$

2 Amicable Numbers 33

and of course 62 is divisible by 31. Note carefully that we now get a different sequence of Ls according to the value of M_p. Thus for M_7 (= 127) we need to compute L_7 using modulo 127. We get

$L_2 = 4$
$L_3 = L_2^2 - 2 = 16 - 2 = 14$
$L_4 = L_3^2 - 2 = 14^2 - 2 = 196 - 2 = 194 \equiv 67 \bmod 127$
$L_5 = L_4^2 - 2 = 67^2 - 2 = 4489 - 2 = 4487 \equiv 42 \bmod 127$
$L_6 = L_5^2 - 2 = 42^2 - 2 = 1764 - 2 = 1762 \equiv 111 \bmod 127$
$L_7 = L_6^2 - 2 = 111^2 - 2 = 12321 - 2 = 12319 = 97 \times 127$

Again L_7 divides exactly by 127 showing that M_7 is prime but we don't have to deal with such horrendous large numbers as before.

In the same way we can show that M_{11} (= 2047) is not prime. We work modulo 2047 and get

$L_2 = 4$
$L_3 = L_2^2 - 2 = 16 - 2 = 14$
$L_4 = L_3^2 - 2 = 14^2 - 2 = 196 - 2 = 194$
$L_5 = L_4^2 - 2 = 194^2 - 2 = 37636 - 2 = 37634 \equiv 788 \bmod 2047$
$L_6 = L_5^2 - 2 = 788^2 - 2 = 620944 - 2 = 620942 \equiv 701 \bmod 2047$
$L_7 = L_6^2 - 2 = 701^2 - 2 = 491401 - 2 = 491399 \equiv 119 \bmod 2047$
$L_8 = L_7^2 - 2 = 119^2 - 2 = 14161 - 2 = 14159 \equiv 1877 \bmod 2047$
$L_9 = L_8^2 - 2 = 1877^2 - 2 = 3523129 - 2 = 3523127 \equiv 240 \bmod 2047$
$L_{10} = L_9^2 - 2 = 240^2 - 2 = 57600 - 2 = 57598 \equiv 282 \bmod 2047$
$L_{11} = L_{10}^2 - 2 = 282^2 - 2 = 79524 - 2 = 79522 \equiv 1736 \bmod 2047$

So we get a remainder of 1736 (luckily the same as when we did it the hard way) and this confirms that 2047 is not prime.

The next Mersenne primes are $M_{13} = 2^{13}-1 = 8191$, $M_{17} = 2^{17}-1 = 131{,}071$ and $M_{19} = 2^{19}-1 = 524{,}287$.

The largest prime number that Mersenne himself correctly predicted was M_{127}. Its value is

$$M_{127} = 170{,}141{,}183{,}460{,}469{,}231{,}731{,}687{,}303{,}715{,}884{,}105{,}727$$

and the corresponding perfect number (the twelfth) is

$$14{,}474{,}011{,}154{,}664{,}524{,}427{,}946{,}373{,}126{,}085{,}988{,}481{,}573{,}677{,}491{,}$$
$$474{,}835{,}889{,}066{,}354{,}349{,}131{,}199{,}152{,}128$$

M_{127} was the largest known prime from its discovery in 1876 by Lucas (confirmation really, since Mersenne predicted it two centuries earlier) until in 1952 (using a computer of course) it was shown that M_{521} was prime. It was quite a big jump since it has nearly four times as many digits as M_{127}.

It has almost always been the case throughout history that the largest known prime is a Mersenne prime. It is easy to see why. The problem with large primes is proving that they are prime. But we have the totally reliable Lucas–Lehmer test in the case of Mersenne numbers and this is easy to apply using modern fast computers.

Sometimes we might know that a number is not prime but yet not know its factors. Thus Lucas showed that M_{67} was not prime in 1876 but the factors were only discovered by the American mathematician Frederick Nelson Cole (1861–1927) in 1903. Cole astounded a meeting of the American Mathematical Society by giving a lecture in which he simply by hand and in silence on one blackboard worked out

$$2^{67}-1 = 147{,}573{,}952{,}589{,}676{,}412{,}927$$

and then on another board worked out

$$193{,}707{,}721 \times 761{,}838{,}257{,}287$$

They were the same. Note that both factors are themselves prime.

What is the point in finding large primes? Until recently it was just for fun like climbing Mt Everest. But now they are used in cryptography as we shall discuss in the last lecture.

Since 1952 many more Mersenne primes have been discovered. At the time of writing the largest known prime number is the 49th known Mersenne prime which was discovered in January 2016. It is $M_{74207281}$ and has 22,338,618 digits. Note that we have to say 49th *known* Mersenne prime since there may be other smaller Mersenne primes that have not yet been discovered. The corresponding 49th known perfect number has twice as many digits.

A little fact that the reader is invited to prove is that the remainder on dividing any perfect number by 3 is always 1 and never 2. This does not apply to the first perfect number which is 6. But it does apply to all the others.

Here is a simple proof that if M_n is prime then n must be prime as well. We use the technique known as *Reductio ad Absurdum* in which we make an assumption and then deduce a contradiction thereby showing that the assumption was false.

Suppose that M_n is prime and that n is not prime but equal to st. Then

$$M_n = 2^n - 1 = 2^{st} - 1 = (2^s)^t - 1$$

Now consider the following giant expression

$$E = (2^s - 1)\{ (2^s)^{t-1} + (2^s)^{t-2} + (2^s)^{t-3} + (2^s)^{t-4} + \cdots + (2^s)^2 + (2^s) + 1 \}$$

If we multiply it out we get

$$E = (2^s)^t + (2^s)^{t-1} + (2^s)^{t-2} + (2^s)^{t-3} + \cdots + (2^s)^3 + (2^s)^2 + (2^s)$$
$$- (2^s)^{t-1} - (2^s)^{t-2} - (2^s)^{t-3} - (2^s)^{t-4} - \cdots - (2^s)^2 - (2^s) - 1$$
$$= (2^s)^t - 1$$

Note how almost all terms cancel in pairs leaving just $(2^s)^t-1$ which of course is 2^n-1 and so E is in fact M_n. However, by construction E is divisible by 2^s-1 and so M_n is not prime. This is a contradiction so n must have been prime.

Amicable numbers

AT LAST we get to the real subject matter of this lecture. An amicable pair of numbers are two numbers, each of which equals the sum of the factors of the other. The smallest pair are 220 and 284. We have

$220 = 2^2.5.11$
$284 = 2^2.71$

The factors of 220 are therefore

1, 2, 4, 5, 10, 11, 20, 22, 44, 55, 110 which add to 284

and the factors of 284 are

1, 2, 4, 71, 142 which add to 220.

We can think of perfect numbers as being amicable with themselves (sounds somewhat introvert).

We recall that defining $\sigma(n)$ to be the sum of the factors of n including n itself then we have the rule that if m and n have no common factors (that is are relatively prime or coprime to each other) then

$$\sigma(m \times n) = \sigma(m) \times \sigma(n)$$

This is useful for working out the sum of factors since we can apply it several times; thus the sum of the factors of 220 is

$$\sigma(220) = \sigma(2^2).\sigma(5).\sigma(11) = (2^3-1).(5+1).(11+1) = 7.6.12 = 504$$

and the sum of the factors of 284 is

$$\sigma(284) = \sigma(2^2).\sigma(71) = (2^3-1).(71+1) = 7.72 = 504$$

(Remember that we often use . rather than × for multiplication in order to save space.)

We see therefore that we can also define an amicable pair of numbers as a pair m and n such that

$$\sigma(m) = \sigma(n) = m + n$$

Pairs of amicable numbers are much more common than perfect numbers. There are no more before we get to 496 which is the next perfect number but then there are four pairs before the next perfect number after that which is 8128. They are (with prime factors in brackets)

1184 ($2^5.37$) and 1210 ($2.5.11^2$)
2620 ($2^2.5.131$) and 2924 ($2^2.17.43$)
5020 ($2^2.5.251$) and 5564 ($2^2.13.107$)
6232 ($2^3.19.41$) and 6368 ($2^5.199$)

Unlike the perfect numbers which all have the same form, the amicable numbers have an extraordinary range of factors with little discernable pattern.

The first pair (220; 284) was known to Pythagoras but curiously enough the next pair (1184; 1210) was only found as recently as 1866 by an Italian schoolboy, Nicolo Paganini. However, two other quite large pairs were known before the middle of the 17th century namely

17,296 ($2^4.23.47$) and 18,416 ($2^4.1151$)
9,363,584 ($2^7.191.383$) and 9,437,056 ($2^7.73,727$)

Amicable numbers are mysterious. In a sense we know all about even perfect numbers. We can generate them from the corresponding Mersenne primes and we have the Lucas–Lehmer test for these. And we suspect that there are no odd perfect numbers. On the other hand we have no known way of generating all amicable pairs.

Most amicable pairs are even. But some are odd, the smallest being

12,285 ($3^3.5.7.13$) and 14,595 (3.5.7.139)

Amicable pairs have factors of various forms. There are usually, but not always, many powers of 2, some small prime numbers and often one or more very large ones.

However, although amicable numbers often have very large primes, the sums of the pairs do not. Thus in the table opposite we see that the largest prime in the sums of the first twelve pairs is 31. But the largest prime in the pairs themselves is never less than 37.

The odd behaviour may be related to a characteristic known as smoothness. A number is smooth if it has only small prime factors. In particular we say that it is 7-smooth if it has no factors larger than 7 and so on. It seems that amicable numbers themselves are somewhat rough but their sums are smoother.

2 Amicable Numbers

m	factors	n	factors	m+n	factors
220	$2^2.5.11$	284	$2^2.71$	504	$2^3.3^2.7$
1184	$2^5.37$	1210	$2.5.11^2$	2394	$2.3^2.7.19$
2620	$2^2.5.131$	2924	$2^2.17.43$	5544	$2^3.3^2.7.11$
5020	$2^2.5.51$	5564	$2^2.13.107$	10584	$2^3.3^3.7^2$
6232	$2^3.19.41$	6368	$2^5.199$	12600	$2^3.3^2.5^2.7$
10744	$2^3.17.79$	10856	$2^3.23.59$	21600	$2^5.3^3.5^2$
12285	$3^3.5.7.13$	14595	$3.5.7.139$	26880	$2^8.3.5.7$
17296	$2^4.23.37$	18416	$2^4.1151$	35712	$2^7.3^2.31$
63020	$2^2.5.23.137$	76084	$2^2.23.827$	139104	$2^5.3^3.7.23$
66928	$2^4.47.89$	66992	$2^4.53.79$	133920	$2^5.3^3.5.31$
67095	$3^3.5.7.71$	71145	$3^3.5.17.31$	138240	$2^{10}.3^3.5$
69615	$3^2.5.7.13.17$	87633	$3^2.7.13.107$	157248	$2^6.3^3.7.13$

The first dozen amicable pairs, their sum and factors.

Note that the sums are often a multiple of 126. Indeed the first five pairs add to multiples of 126 thus

$220 + 284 = 504 = 126 \times 4$
$1184 + 1210 = 2394 = 126 \times 19$
$2620 + 2924 = 5544 = 126 \times 44$
$5020 + 5564 = 10584 = 126 \times 84$
$6232 + 6368 = 12600 = 126 \times 100$

But it doesn't work for the next pair which are 10744 and 10856 since their sum is $21600 = 2^5.3^3.5^2$. Note that $126 = 7 \times 18$ and 7 is not a factor of 21600.

Another curious thing is that since perfect numbers can be seen as amicable with themselves one might think that double a perfect number might exhibit some degree of smoothness. But this is not so since a perfect number is always a power of 2 multiplied by a Mersenne prime and is decidedly rough.

Although we know no way of generating all amicable numbers, there are formulae that generate some pairs. The Arabic mathematician Thabit ibn Qurra (824–901) showed that if

$p = 3.2^{n-1} - 1$
$q = 3.2^n - 1$
$r = 9.2^{2n-1} - 1$

are all prime then the pair

$M(2^n.p.q)$ and $N(2^n.r)$

are always amicable.

It's quite amazing that Thabit discovered this so early. If we try $n = 2$, then $p = 5$, $q = 11$, $r = 71$ and these are all prime and this gives the first pair (220; 284).

The case $n = 4$ gives $p = 23$, $q = 47$, $r = 1151$ and thus the pair (17,296; 18,416). This was discovered in the early 14th century. The case $n = 7$ gives $p = 191$, $q = 383$, $r = 73,727$ and thus the pair (9,363,584; 9,437,056). This was discovered in the 17th century. Recently it has been shown by brute force that no other values of n less than 20,000 give amicable pairs.

Lots of things are not known about amicable pairs. For example, it is not known whether

any amicable pairs are relatively prime to each other,

any amicable pairs are one odd and one even,

any amicable pairs have only one divisible by 3,

any amicable pairs have all factors greater than 5.

But it is known that some odd amicable pairs are not divisible by 3. And it is known that even amicable pairs never have a factor of 3 (that's very strange).

Amicable multiplets

IN A SIMILAR WAY we can also define amicable triplets, amicable quadruplets and indeed amicable multiplets. By analogy with the formula

$$\sigma(m) = \sigma(n) = m + n$$

we can define an amicable triplet as three numbers l, m, n such that

$$\sigma(l) = \sigma(m) = \sigma(n) = l + m + n$$

There are many examples of such triplets such as

1980;   2016;   2556
9180;   9504;   11,556
21,168;   22,200;   27,312

We can similarly define amicable multiplets such as the quadruplet

554,130,720 ($2^5.3^3.5.11.13.23.29$),
444,169,440 ($2^5.3^3.5.11.13.719$),
481,546,080 ($2^5.3^3.5.17.79.83$),
491,153,760 ($2^5.3^3.5.41.47.59$)

Such multiplets up to order 7 are known.

Sociable cycles

ANOTHER GROUPING is to consider cycles. Is it possible to find a cycle of numbers such that each is the sum of the factors of the previous one and for the first to be the sum of the factors of the last? Yes it is and these are called sociable cycles. Amicable numbers are of course simply two-cycles and perfect numbers are one-cycles.

The history of sociable cycles is quite brief. The first two sociable cycles were discovered in 1910 by the Belgian mathematician Paul Poulet (1887–1946). They involve the smallest numbers of any sociable cycles. They are a 5-cycle and a 28-cycle. Thus the multiple cycle with the smallest numbers is the 5-cycle.

12,496;   14,288;   15,472;   14,536;   14,264;   and then 12,496 again

and the next cycle is an amazing 28-cycle thus

14,316;   19,116;   31,704;   47,616;   83,328;   177,792;   295,488;
629,072;   589,786;   294,896;   358,336;   418,904;   366,556;   274,924;
275,444;   243,760;   376,736;   381,028;   285,778;   152,990;   122,410;
97,946;   48,976;   45,946;   22,976;   22,744;   19,916;   17,716

What is very surprising is that despite the 28-cycle being one of the first two sociable cycles ever discovered, no longer cycle or indeed any cycle half as long has been found in the more than 100 years that have followed.

Over a hundred 4-cycles are known, the first few are

1,264,460;   1,547,860;   1,727,636;   1,305,184
2,115,324;   3,317,740;   3,649,556;   2,797,612
2,784,580;   3,265,940;   3,707,572;   3,370,604
4,938,136;   5,753,864;   5,504,056;   5,423,384
7,169,104;   7,538,660;   8,292,568;   7,520,432

Although there are many 4-cycles, no 3-cycles are known, and it is thought, although it has not been proved, that maybe there are no 3-cycles.

At the time of writing, other cycles known are five 6-cycles, three 8-cycles and one 9-cycle. Their smallest numbers are

21,548,919,483	start of 6-cycle
90,632,826,380	start of 6-cycle
1,771,414,411,016	start of 6-cycle
3,524,434,872,392	start of 6-cycle
4,773,123,705,616	start of 6-cycle
1,095,447,416	start of 8-cycle
1,276,254,780	start of 8-cycle
7,914,374,573,864	start of 8-cycle
805,984,760	start of 9-cycle

Just as for amicable pairs all members of known cycles have the same parity – that is either they are all even or all odd. Most cycles are even but one of the 6-cycles is odd and several 4-cycles are odd.

The cycles are intriguing. Having found the 5-cycle and 28-cycle with fairly small numbers, one might expect lots of other cycles of many different lengths. It is disappointing therefore that most of the rest are 4-cycles. There is clearly scope for the reader to look for some others.

One thing that is different from amicable pairs is that whereas no even amicable pair ever has 3 as a factor, this does not apply to sociable cycles. Many of the numbers in the 28-cycle are divisible by 3 (such as $14316 = 2^2.3.1193$).

The fact that there are no 3-cycles is surprising but perhaps reflects the fact that two is company but three is a crowd.

Fermat numbers

PIERRE DE FERMAT (1601–1665) was a lawyer and magistrate and also a brilliant "amateur" mathematician. He didn't bother about winning fame but did mathematics for the fun of it. As a consequence some of his work has come down to us in a sketchy format. The most famous of course is his Last Theorem which says that the equation

$$x^n + y^n = z^n$$

where x, y, z, and n are integers does not have any solutions for n greater than 2. Of course if n is 2 then there are a host of solutions corresponding to Pythagoras such as $3^2 + 4^2 = 5^2$. Fermat claimed to have found a proof but that the margin of his paper was too small to give it. It was only proved very recently by Andrew Wiles in 1995.

Fermat was interested in numbers of the form

$$F_n = 2^{2^n} + 1$$

which are somewhat similar to the Mersenne numbers $M_n = 2^n - 1$ discussed earlier. In the case of the Fermat numbers the first few are

$$F_1 = 2^2+1 = 5, \ F_2 = 2^4+1 = 17, \ F_3 = 2^8+1 = 257, \ F_4 = 2^{16}+1 = 65{,}537$$

and these are all prime. Fermat accordingly conjectured that all Fermat numbers are prime. However, this is not the case, as was shown by Euler who found in 1732 that

$$F_5 = 2^{32} + 1 = 4{,}294{,}967{,}297 = 641 \times 6{,}700{,}417$$

Observe that 6,700,417 is indeed prime. It is believed although it has not been proved that all Fermat numbers after F_4 are composite.

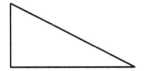

A right-angled triangle with sides 1, 2, and √5.

An interesting fact regarding Fermat primes is that a regular polygon can be constructed using ruler and compasses alone if the number of its sides is a Fermat prime. So a 17-gon can in principle be constructed although measurement errors make it somewhat tricky.

The basic reason why the constructions can be done is that we have to solve an equation of degree $F-1$. In the case of the 17-gon we thus have to solve an equation of degree 16. It turns out that this can be done by solving a series of nested quadratic equations since $16 = 2^4$. The roots of a quadratic equation just involve square roots and these can be found by ruler and compasses alone.

In the case of a pentagon we need to find the square root of 5 and this is easily done since the hypotenuse of a right-angled triangle whose other sides are 1 and 2 is of course √5 using Pythagoras as shown above.

From this we can construct the Golden Number $\tau = (\sqrt{5} + 1)/2$. Now the diagonal of a pentagon of side 1 is simply $2\tau + 1$ and hence we can then construct the pentagon itself. See for example *Gems of Geometry* by the author.

In the case of the 17-gon we need to find the value of a nasty expression involving the square root of 17. See the references for details.

Fibonacci numbers

IN 1202, LEONARDO OF PISA (c 1175–1250) who was commonly called Fibonacci, meaning son of good fortune, produced a book entitled *Liber Abbaci*. One topic concerned the breeding of rabbits and in particular introduced a series of numbers now called Fibonacci numbers. It is a series in which each number is the sum of the previous two and the first two are both one. So the series begins

1, 1, 2, 3, 5, 8, 13, 21, 34, 55, 89, ...

An important property of the series is that the ratios of successive pairs are closer and closer approximations to the golden number τ just mentioned in connection with Fermat numbers. The ratios are

1/1 = 1; 2/1 = 2; 3/2 = 1.5; 5/3 = 1.666...; 8/5 = 1.6; 13/8 = 1.625;

and so on.

The golden number τ is 1.618033989.... It has many neat properties such as

$$\tau - 1 = 1/\tau \quad \text{and} \quad \tau + 1 = \tau^2.$$

We will encounter the Fibonacci numbers again in Lecture 4 on Fractions when we discuss Egyptian fractions and continued fractions, and in Lecture 8 on Primes when we discuss the greatest common divisor. They also turn up in Appendix B and Appendix G. In particular, Appendix G gives an explicit formula for the *n*th member of the series.

Further reading

A COMPREHENSIVE BOOK on this topic is *Perfect, Amicable and Sociable Numbers* by Song Yan – this is not for the faint-hearted. Various websites are interesting. See for example http://mathworld.com/SociableNumbers.html. A list of all known cycles will be found at http://djm.cc/sociable.txt. The topic of smoothness is discussed at http://en.wikipedia.org/wiki/Smooth_number.

The Mersenne, Fermat, and Fibonacci numbers are discussed in *Elementary Number Theory* by David M Burton and also in *Elementary Number Theory and its Applications* by Kenneth E Rosen.

The construction of polygons with a Fermat prime number of sides is discussed in *Introduction to Geometry* by Coxeter and also in *Makers of Mathematics* by Hollingdale. Coxeter gives the explicit construction for the 17-gon and Hollingdale outlines the nasty process of solving the nested quadratic equations.

Exercises

1. Show that $12496 = 2^4.11.71$. Hence show that its factors (all of them, not just the prime ones) add to 14288. Then find the factors of 14288 (its largest prime factor is 47) and show that they add to 15472. And similarly show that the factors of 15472 (largest prime is 967) add to 14536 and that those of 14536 (largest prime 79) add to 14264. Finally, show that the factors of 14264 (largest prime 1783) add to 12496 thus completing the cycle of 5.

2. Show that the remainder on dividing an even perfect number (other than 6) by 3 is always 1.

3 Probability

IN THIS LECTURE we look at some of the basic ideas of probability theory. This has exciting applications in gambling which are inevitably misplaced. We start by considering tossing coins and throwing dice and then look at some unexpected examples where our intuition breaks down completely.

Heads or tails

ONE of the most important ideas to grasp is that events such as tossing a coin are independent of each other; we say that they are unrelated events. It is clear that if I toss an unbiased coin then the probabilities of it landing heads or tails are equal. We say that each event has probability of 0.5.

If I toss a coin six times and it comes down heads each time then the question arises as to what is the probability of it coming down heads on the seventh occasion. The answer is that it is just 0.5 still. The coin has no memory and it doesn't care that it came down heads for the last six times. There is no Law of Chance that says your luck must change sometime.

It is indeed unlikely that a coin should come down heads six times in a row. If I start the experiment and declare I am now going to toss my coin six times and ask what is the probability that it comes down heads each of the six times then the six events are unrelated and the probability of them all occurring is obtained by multiplying the individual probabilities together. So we get

$$0.5 \times 0.5 \times 0.5 \times 0.5 \times 0.5 \times 0.5 \times 0.5 = 1/64$$

What this means is that if I do the experiment lots of times then on average I will get 6 heads about 1 in every 64 times I do it.

Suppose I toss six coins together. We assume that the coins do not interfere with each other in mid-flight. Each coin will behave independently and the probability of each coin landing heads will be 0.5. So again the probability of all six landing heads together is just 1/64.

Suppose we toss just two coins. With equal probability each will be H or T. There will be four possible results

HH, HT, TH, TT

and each will occur with probability 1/4. So (assuming that the coins are not marked so that we cannot distinguish them), we see that there are three possible outcomes

two heads	1 way	prob 1/4
one head and one tail	2 ways	prob 1/2
two tails	1 way	prob 1/4

So there is an evens chance that both will be the same. Note that we cannot argue this in a vague way by saying that they will either be the same or not the same, so it's 50–50; we have to consider the details. But it is important to note that the probabilities add to 1.0.

Now consider three coins. This time there are 8 equally possible outcomes

HHH, HHT, HTH, HTT, THH, THT, TTH, TTT

and this gives

three heads	1 way	prob 1/8
two heads and one tail	3 ways	prob 3/8
one head and two tails	3 ways	prob 3/8
three tails	1 way	prob 1/8

And similarly with four coins we get

four heads	1 way	prob 1/16
three heads and one tail	4 ways	prob 4/16 = 1/4
two heads and two tails	6 ways	prob 6/16 = 3/8
one head and three tails	4 ways	prob 4/16 = 1/4
four tails	1 way	prob 1/16

If we do it with five coins then the probabilities are 1/32, 5/32, 10/32, 10/32, 5/32, 1/32. And with six coins they are 1/64, 6/64, 15/64, 20/64, 15/64, 6/64, 1/64. We can represent this by a barchart or histogram as shown below. The word histogram is derived from the Greek ιστος (*histos*) meaning a mast.

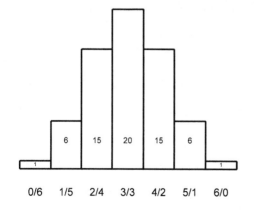

Histogram showing the probabilities when 6 coins are tossed. Note the symmetry.

1/5 = 1 tail + 5 heads.

3 Probability

We see a pattern emerging. In fact the groups of numbers 1,2,1 – 1,3,3,1 – 1,4,6,4,1 are the rows of Pascal's triangle which can be written thus

```
                    1
                  1   1
                1   2   1
              1   3   3   1
            1   4   6   4   1
          1   5  10  10   5   1
        1   6  15  20  15   6   1
      1   7  21  35  35  21   7   1
    1   8  28  56  70  56  28   8   1
```

Each number is the sum of the two adjacent numbers in the row above. The numbers in each row add up to successive powers of 2 thus 2, 4, 8, 16, 32. We might also recognise the numbers as the digits of successive powers of 11, thus $11^2 = 121$, $11^3 = 1331$, $11^4 = 14641$. After that, carrying obscures the sequence so that $11^5 = 161051$ is not exciting.

The triangle is named after Blaise Pascal (1623–1662), the famous French mathematician who explored it in detail and established the basic ideas of probability theory.

More abstractly the numbers in Pascal's triangle are the coefficients in the expansion of $(x + y)^n$. Thus

$(x + y)^2 \quad = \quad x^2 + 2xy + y^2$

$(x + y)^3 \quad = \quad x^3 + 3x^2y + 3xy^2 + y^3$

$(x + y)^4 \quad = \quad x^4 + 4x^3y + 6x^2y^2 + 4xy^3 + y^4$

and so on. They are called the Binomial coefficients and the probability distribution is hence called the Binomial distribution. The general formula for the number of ways of choosing r things from n things is denoted by ^nC_r and is

$$\frac{n!}{(n-r)! \times r!} \quad \text{where } n! = n \times (n-1) \times (n-2) \times (n-3) \times \cdots \times 2 \times 1$$

So for example, putting $n = 5$ and $r = 2$ we get

$$\frac{5 \times 4 \times 3 \times 2 \times 1}{(3 \times 2 \times 1) \times (2 \times 1)} = \frac{5 \times 4}{2} = 10$$

which is the correct answer for the numbers of ways of getting 2 heads when tossing 5 coins. And so the probability of getting 2 heads when tossing 5 coins is 10/32 since there are $2^5 = 32$ possible outcomes.

Note carefully that this does not mean that if we toss 5 coins exactly 32 times then they will come down with 2 heads exactly 10 times. It is just that it will be 10 on average.

We can somewhat laboriously work out the probability that it will be exactly 10 times. For this to happen then they must come down 10 times with 2 heads and 22 times they must not. Let us suppose first of all that it is in fact the first 10 attempts that have 2 heads and the remainder that do not.

The probability of getting 2 heads is 10/32 and the probability of not getting 2 heads is 22/32. So the probability of this overall sequence is

$$(10/32)^{10} \times (22/32)^{22}$$

This is rather small – about 0.000000002. But the order in which we get the 10 pairs of heads doesn't matter. And in fact we can choose the 10 successful tosses in $32!/(10! \times 22!)$ ways (using the magic formula above) so we have to multiply by this as well. The result is about 0.151 which is perhaps surprisingly low. We can similarly work out the probability that 2 heads will happen only 9 times and so on. The results are shown in the table below.

These probabilities add up to 1 as they should do. So the message is that probabilities are just that and not prescriptive certainties. Although the average number of times that 2 heads occurs is indeed 10 in 32 tosses, and the actual value is more likely to be 10 than any other value, nevertheless values anywhere between 5 and 15 are not surprising.

We can illustrate the range of possible outcomes using a histogram as shown opposite. The individual probabilities are again represented by the height of a column. The highest column corresponds to the most likely value which is 10.

no of times	probability	no of times	probability
0	0.0000	13	0.0762
1	0.0001	14	0.0470
2	0.0006	15	0.0256
3	0.0029	16	0.0124
4	0.0095	17	0.0053
5	0.0242	18	0.0020
6	0.0495	19	0.0007
7	0.0837	20	0.0002
8	0.1189	21	0.0001
9	0.1442	22	0.0000
10	0.1507	23	0.0000
11	0.1370	...	...
12	0.1090	32	0.0000

Probabilities of two heads a given number of times when tossing 5 coins 32 times.

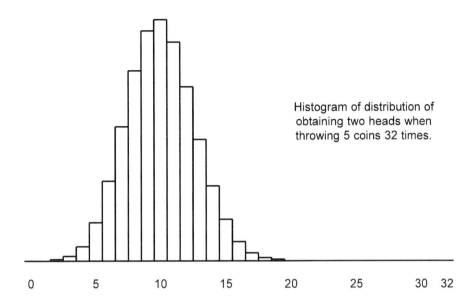

Histogram of distribution of obtaining two heads when throwing 5 coins 32 times.

Distributions

WE HAVE JUST LOOKED at the distribution of getting 2 heads when tossing 5 coins 32 times. It has various properties. The expected value or *mean* is 10. However, it is skew and not symmetric. The most likely value is 10, but the probabilities of getting 9 or 11 are not quite the same. It is possible (although extremely unlikely) to get a score of 22 (that is 10 plus 12) but clearly impossible to get a score of 10 minus 12.

Some distributions are symmetric. For example the distribution of heads and tails when tossing 6 coins several times is completely symmetric.

An interesting property of a distribution is the *standard deviation*. This is a measure of how far off the mean a single measurement will be on average. In the case of tossing 5 coins, the average number of times we get 2 heads is 10 but as we have seen a typical result will not be 10. So how far off will the result be on average?

If the mean is m, and a typical result is x, then we define the *variance* as the average of

$$(m - x)^2$$

(the square ensures that this is positive whether x is less than or greater than the mean) and then we take the square root to obtain the standard deviation.

In the case of the Binomial distribution the variance can be shown to be

(number of trials, n) × (prob of success, p) × (prob of failure, $1-p$)

Applying this to the coin tossing, the number of trials is 32, the probability of success is 10/32 and the probability of failure is 22/32. The final result is that the variance is 220/32 = 6.875. The standard deviation is then the square root of this which is 2.62.... So on average the result will deviate from 10 by this amount which looks reasonable from the histogram above. As a rule of thumb we should not be surprised with a trial that gives a result up to twice the standard deviation from the mean.

It is important to be aware of the difference between probability and statistics.

Probability is about predicting outcomes of trials given knowledge of the probabilities involved. Given a perfect coin, we might ask what is the probability of getting two heads in succession?

Statistics is the opposite. We do a number of trials and from these we obtain estimates of the probabilities involved. The more trials we do the better our estimates should be. We can find the *likelihood* that the coin is perfect within some bounds – perhaps 1%. We do not say probability because the coin is either perfect or not – there is no chance involved.

For example, in the coin tossing if we do 32 trials then we get an estimate of the underlying probability by dividing the number of successes by 32. As we have seen we are likely to get a value in the range of 10 plus or minus 2.62 divided by 32 which is a range of 0.23 to 0.39. (Hmm – pretty vague.) If we do 128 trials, then the mean is 40 and the standard deviation is doubled to 5.24.... The result is that the range becomes 0.27 to 0.35. So if we do four times as many trials then the range is halved.

The author is at risk of being a bit slipshod here. If we are dealing with known probabilities with mean m then the variance is obtained by the average of $(m-x)^2$ as mentioned above. However, if we are dealing with statistics and seeking estimates of the mean and variance where the underlying probabilities are not known and we do n trials, then we obtain an estimate of the mean as we expect by adding the trial values x and dividing by n. We will denote this estimate by m' in order to distinguish it from the true underlying mean m which we do not know.

However, in estimating the variance from the trial values we divide the sum of the various $(m'-x)^2$ by $(n-1)$ and not by n. This is because we are using the estimate m' rather than the true value m; in essence we lose a value because we do not know m. For example, from just one trial we get a (pretty rough) estimate m' for the underlying m. But clearly we can have no estimate whatsoever of the variance from a single trial. In practice, if we are doing lots of trials, then this subtlety does not matter.

Clearly, we have to do a lot of trials to get a really good estimate. But in general if we do more trials then we get a better estimate. This in essence is the Law of Large Numbers.

3 Probability 49

Shake, rattle, and roll

DICE ARE the other classic gaming mechanism after coins. In the case of a coin there are two possibilities H and T and these occur with probability 1/2. In the case of a 6-sided die, there are six possibilities and these occur with equal probability of 1/6.

Dice are such that the opposite faces almost always add to 7. Thus the opposite faces are usually (1, 6), (2, 5), and (3, 4). Even Roman dice show this regularity but sometimes one encounters rogue examples that do not conform. However, even if they do conform then they come in various versions. If the 6 is on top and the 5 facing, then the 4 can be on the left or right. So these are mirror images. But there are other variations. The 3 can slope from bottom left to top right or the other way. And the same applies to the 2. Moreover, the 6 can be made as 2×3 or as 3×2. So that gives 16 possible variations as shown below.

Style 4 is perhaps the most common. Also shown below are three real dice probably manufactured as part of the same batch. Clearly the process ensures that the opposite faces add to seven but imposes no other constraints since we find examples of styles 4, 10, and 11.

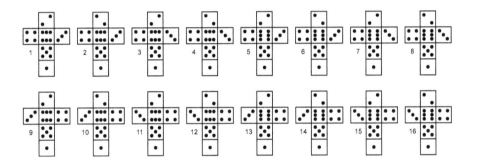

Sixteen forms of nice dice.

Samples of actual dice showing styles 4, 10, and 11.

50 Nice Numbers

A B	1	2	3	4	5	6
1	2	3	4	5	6	7
2	3	4	5	6	7	8
3	4	5	6	7	8	9
4	5	6	7	8	9	10
5	6	7	8	9	10	11
6	7	8	9	10	11	12

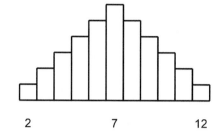

Distribution of score when throwing two dice.

If we roll two dice A and B then the various total scores do not occur with equal probability. Rolling a double six giving a score of 12 only occurs if both dice come up with six and this occurs with probability $1/6 \times 1/6 = 1/36$ (remember that we multiply probabilities when combining unrelated events in this way).

Rolling a score of 4 can occur in 3 different ways, namely

1,3 or 2,2 or 3,1

and so the probability of a score of 4 is $1/36 + 1/36 + 1/36$ (we add the probabilities when the events are combined with "or") = 1/12. Similarly, a score of 7 can occur in 6 different ways, namely

1,6 or 2,5 or 3,4 or 4,3 or 5,2 or 6,1

and so the probability of a score of 7 is $6/36 = 1/6$.

The various probabilities can be represented visually using a table or more traditionally using a histogram as shown above. The table instantly reveals why 7 (in blue) is the most likely score and that 12 (in green) is unlikely.

The histogram clearly reveals that the distribution is completely symmetric with mean, m, of 7. We can compute the standard deviation quite easily. We consider each possible score x, subtract it from the mean and then square it to give v. We then take the average of these taking due account of the fact that the scores occur with different probabilities. This gives the variance and we then take the square root to obtain the standard distribution. So we get

x	p = prob of x	$v = (m-x)^2$	$v \times p$
2, 12	1/36	25	25/36
3, 11	2/36	16	32/36
4, 10	3/36	9	27/36
5, 9	4/36	4	16/36
6, 8	5/36	1	5/36
7	6/36	0	0

The sum of all the $v \times p$ (taking account of the duplications) is $210/36 = 5.833...$, so the standard deviation is the square root of this which is $2.415...$.

We can now go on to consider the probability of getting a particular score a certain number of times in a sequence of throws just as we did for tossing coins. We might ask for the distribution of how many times we get 7 when throwing two dice 12 times. We know that the probability of getting 7 is 1/6 so on average we might expect to get 7 just twice. We use exactly the same technique as with the coins.

The probability of not getting 7 at all in 12 throws is

$$(5/6)^{12} = 0.112...$$

because the probability of not getting 7 on a throw is 5/6 and this has to happen each of the 12 times.

The probability of getting 7 just once in 12 throws is

$$(1/6)^1 \times (5/6)^{11} \times 12 = 0.269...$$

because on one throw we need 7 and on 11 throws we must not have 7. Moreover, the successful throw can be chosen in 12 ways from the throws and so we have to multiply by 12 as well.

The probability of getting 7 just twice in 12 throws is

$$(1/6)^2 \times (5/6)^{10} \times (12 \times 11/2) = 0.296...$$

where we have used the general formula given earlier for choosing the 2 successful throws from the 12 throws.

We get a table of results as follows

no of times	probability
0	0.112
1	0.269
2	0.296
3	0.197
4	0.0888
5	0.0284
6	0.00663
7	0.00113
8	0.000142
9	0.0000126
10	0.000000758
11	0.0000000275
12	0.000000000459

The most likely number of times is 2 as expected. But 1 and 3 are also quite likely.

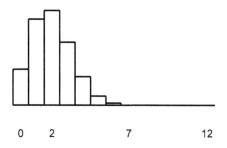

Distribution of number of times we get 7 when throwing two dice 12 times.

This distributions is rather skew as is clearly revealed by the histogram shown above. The variance is $12 \times 1/6 \times 5/6 = 60/36$ which is 1.666... so the standard deviation is the square root of this which is 1.29.... This ties in well with the fact that 1 and 3 are also quite likely.

The normal distribution

MOST NATURALLY OCCURRING distributions approximate to a sort of bell shape with the expected outcome being most likely and others falling off rapidly either way. We have seen that the Binomial distribution is of this form. It is interesting to look at the Binomial distribution for various values of the parameters in some detail.

Earlier we showed a histogram of the distribution of the result of tossing 6 coins. It is shown again below in a slightly smaller scale. Opposite are shown the histograms for tossing 24 and 96 coins respectively in comparative scales. Indeed they are such that the total areas are exactly the same (6400 sq pts) and the standard deviations are very similar. However, the extreme ends of the distributions become extreme indeed and are not shown. With 6 coins the probability of zero tails is 1/64; with 96 coins it is $1/2^{96}$. Now 2^{96} is about 10^{33} which is the sort of number that astronomers discuss. The probability of getting

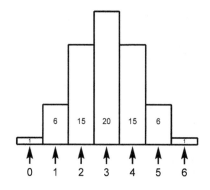

Histogram showing the probabilities when 6 coins are tossed. The numbers below the bars show the number of tails.

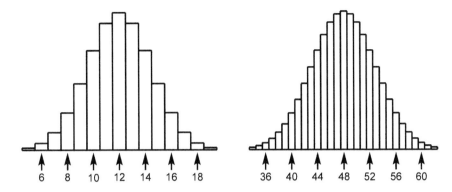

Histograms showing the probabilities when 24 (left) and 96 (right) coins are tossed.

exactly equal numbers of heads and tails also diminishes but not so violently; it is 20/64 = 0.3125 for 6 coins; about 0.1612 for 24 coins; and about 0.08122 for 96 coins – roughly halving each time.

But the overall shape is clearly converging. The heights of the central bar in pts are 100.0 (6 coins), 103.154 (24 coins), 103.963 (96 coins), 104.166 (384 coins). Similarly, the comparable heights of the lower corresponding bars are 75.0, 74.185, 74.772, 74.696 and 30.0, 28.056, 27.617, 27.514. Indeed as the number of coins gets very large the distribution approaches the important mathematical distribution known as the normal distribution.

The normal distribution to the same scale is shown below. The equation is

$$y = e^{-x^2/2} / \sqrt{2\pi}$$

which at first sight is somewhat gruesome. The number e is the base of natural logarithms; its value is 2.71828.... The peak is when x is zero and so is $1/\sqrt{2\pi}$ which is 0.3989.... We will encounter e again in Lecture 8 when discussing complex numbers.

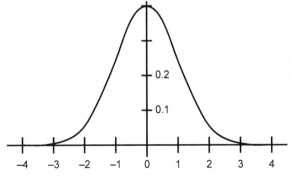

The normal distribution. Note that the vertical and horizontal scales are not the same.

Note that the values tail off very rapidly indeed. When $x = 0$, $y = 0.39894$; when $x = 1$, $y = 0.24197$; when $x = 2$, $y = 0.05399$; when $x = 3$, $y = 0.00443$.

The standard deviation is exactly 1.0. About 68% of the area is from $x = -1.0$ to $x = +1.0$ and 95% is between -2.0 and $+2.0$. Hence the rule that one should not be surprised if individual results are twice the standard deviation from the mean.

Another interesting feature of the curve is that the points of inflexion are at $x = -1.0$ and $+1.0$. The points of inflexion are where the curve changes from being convex to concave.

In order to see how the binomial distribution approaches the normal distribution, it is convenient to use the rather surprising approximation for $n!$ discovered by the Scottish mathematician James Stirling (1692–1770). The full approximation uses a series but for most purposes the first term is enough. It is

$$n! \approx \sqrt{(2\pi n)} \times (n^n/e^n)$$

This formula is surprisingly accurate for large n and not too bad for modest values. For example taking $n = 6$ we have $6! = 720$; Stirling gives $710.078...$ which is in error by about 1.4%. For $n = 12$, $12! = 479,001,600$; Stirling gives $475,687,486$ which is in error by about 0.7%.

The peak value of the binomial distribution (for even n) is $^nC_{n/2}$ which using the formula given at the beginning of this lecture is $n!/((n/2)!)^2$. Using Stirling's approximation and after much cancellation this becomes $2^n \times \sqrt{(2/\pi n)}$. In the case of $n = 6$ this is about 20.847 (the true value is 20) and when $n = 24$ it is about 2,732,463 (the true value is 2,704,156) and when $n = 96$ it is $6.451... \times 10^{27}$ (the true value is 6,435,067,013,866,298,908,421,603,100). Enough!

Biological distributions such as the distribution of the height of a group of people typically approximate to the normal distribution. Of course the distribution has to be adjusted to suit the scale of values concerned.

An abnormal distribution

AN IMPORTANT PROPERTY of most distributions is that they have a well-defined mean and standard distribution. Although there is no Law of Chance that says that one's luck must change, there is a Law of Large Numbers that says roughly that things turn out as expected on average and that the more you try something the better the estimate. In other words experimentally derived results converge and give a good approximation with a reasonable number of trials.

For example consider tossing a coin 10 times; we expect the mean number of heads to be 5 and the variance to be (using the formula given earlier) $10 \times 0.5 \times 0.5 = 2.5$, so that the standard deviation is the square root of this which is about 1.58. With 100 tosses we expect the mean to be 50 and the standard deviation to be 5 and so on. If we scale the means to be 0.5 throughout then we expect the standard deviations to be 0.158 for 10 tosses and 0.05 for 100 tosses.

Tosses	T1	T2	T3	T4	T5	T6	Mean	SD
10	0.6000	1.0000	0.6000	0.8000	0.1000	0.5000	0.6000	0.2944
100	0.5200	0.5000	0.4800	0.4900	0.4900	0.4700	0.4917	0.0178
1000	0.5290	0.5170	0.5110	0.4860	0.5020	0.4860	0.5052	0.0166
10000	0.4959	0.5012	0.4926	0.4975	0.5000	0.5064	0.4989	0.0045
100000	0.5015	0.5000	0.5021	0.5003	0.5015	0.4990	0.5007	0.0013
1000000	0.5007	0.4997	0.4997	0.4994	0.5001	0.5009	0.5001	0.0005

Results of tossing a coin many times.

The table above shows the results from simulating the tossing of a coin 10, 100, ... 1,000,000 times and doing each experiment 6 times. The more tosses are performed the closer the average of 6 trials comes to the hoped-for 0.5. Also the standard deviations progressively reduce as expected.

Incidentally, note that the author did not toss an actual coin 10 times and do that experiment 6 times. And he certainly did not do it a million times. The experiment was simulated using a pseudo-random number generator on a computer.

But there is one distribution that is rather peculiar. It is known as the Cauchy distribution after Augustin-Louis Cauchy (1789–1857) the French mathematician who was a key figure in the development of the theory of complex numbers.

Suppose we take a point P at unit distance from a line and draw another line at random through the point. It will (unless by a fluke it is parallel) cut the first line in a point Q at distance x from the point O on the line nearest to P. What is the distribution of x? We can view the picture as below. It's a bit like throwing a semi-infinite stick with one end fixed.

We can assume that the angle θ lies uniformly at random between $-90°$ and $+90°$. The value of x is then simply $\tan \theta$. By symmetry the mean value of x must surely be zero. Note that x can range from minus infinity to plus infinity. Now what is the standard deviation of x?

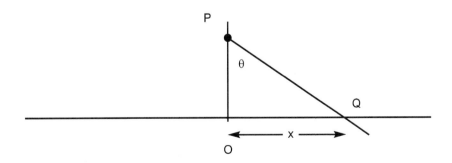

A strange distribution.

56 Nice Numbers

Throws	T1	T2	T3	T4	T5	T6	Mean	SD
10	-2.670	-2.595	1.451	44.765	0.932	0.041	6.987	18.352
100	-0.591	-1.638	1.032	1.431	-0.034	-11.405	-1.868	4.765
1000	-0.784	0.147	-1.158	9.208	-0.163	-35.488	-4.707	14.979
10000	-0.194	-0.903	4.755	0.634	-2.614	0.451	0.355	2.269
100000	0.754	18.512	-0.400	0.312	-0.479	-1.407	2.882	7.591
1000000	1.034	0.608	-0.990	0.751	0.514	-3.320	-0.234	1.542

Results of throwing a stick many times.

The answer very surprisingly is that the standard deviation is not defined. Moreover, the mean itself is not defined even though by symmetry one would expect it to be zero. This means that if we draw many such lines (or throw sticks) and note the values x then no matter how many times we do it the average is not guaranteed to approach zero! The Law of Large Numbers does not apply to this strange distribution.

We can do a simulation using a pseudo-random number generator as done for the coin tossing. Some typical results are shown above. It is clear that the numbers are not converging. Thus on one of the trials of throwing the stick a thousand times the mean for those trials was –35.

Tossing for π

ANOTHER STICK TOSSING problem is better behaved. Suppose we have a ruled grid of parallel lines 1 unit apart. Now drop a rod of unit length at random onto the grid. What is the probability that it will cross a line?

The answer is $2/\pi$. So here is an unlikely way of estimating π. The proof of this uses just elementary calculus. (Nervous readers can skip the next bit.)

It is simpler to consider the rod to be 2 units long and for the lines to be 2 units apart. Now consider the midpoint of the rod and consider the line nearest to this midpoint when it falls. Suppose the distance to the line is x and that the rod is at an angle θ as shown. The rod will meet the line provided that x is less

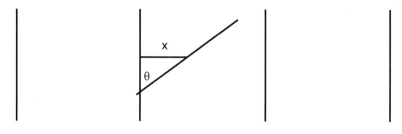

A grid of parallel lines.

than sin θ. Moreover, x is uniformly distributed between 0 and 1. So for a given θ the probability that the rod will meet the line is simply

sin θ

As a check, note that if θ is 0 then sin θ is zero and the stick never meets a line whereas if θ is 90° then sin θ is 1 and the stick is bound to meet a line.

We can also assume that the angle θ is uniformly distributed between 0° and 180° but of course we have to use radians so that becomes 0 and π. Hence the overall probability is obtained by integrating this and we get

$$p = \frac{1}{\pi} \times \int_0^\pi \sin \theta \, d\theta = \frac{1}{\pi} \times \left[-\cos \theta \right]_0^\pi = \frac{2}{\pi}$$

The term $1/\pi$ is because the range of integration is from 0 to π and then we perhaps remember that the integral of sin x is −cos x so that the integral part in square brackets is −cos π + cos 0 which is 1+1 = 2.

And so finally we find that

$$p = 2 / \pi$$

as promised.

Is this an easy way to estimate π? If we do 1000 trials then the mean of the number of successes will be 2000/π (about 636) and the variance will be $n \times p \times (1-p)$ which is about 231, so the standard deviation will be about 15. So allowing two standard deviations either way we might expect to get π to about 1 part in 20.

It's not a good way to find π. It's made worse by it not always being clear whether the rod is cutting a line or not. Incidentally, this exercise was first proposed by the French naturalist and mathematician, the Comte de Buffon (1707–1788) and is consequently often called Buffon's Needle.

Double or quits

A RENOWNED PLOY for winning is to double your stake to cover losses. Suppose we are playing a game with evens chances of winning.

We stake £1. If we win, we just walk away with a gain of £1. If we lose we stake £2 the next time. If we win then we have covered our previous loss and are up with a gain of £1 and then we walk away. If we lose we stake £4 and so on.

Clearly we must win eventually and then we have made a gain of £1. We can then repeat the process all evening. Some runs may be long but all must eventually result in a gain of £1.

What is the flaw? The problem is that there will eventually be a run of losses which uses up all our capital so that we cannot continue and that will wipe out all our gains.

A medical problem

MEDICAL STATISTICS are easily misunderstood. Suppose there is a test for some disease (the nadgers say) which is 99% accurate. That seems very reliable. Suppose also that the nadgers is a reasonably common condition and occurs in 1 person in 1000 of the population.

You are tested for the nadgers and worryingly the test is positive. What is the probability that you do indeed have the nadgers? Should you have the treatment which we can assume is nasty?

One's first reaction is that since the test is so reliable it is almost certainly 99% likely that you have the dreaded nadgers and so need the treatment.

But it is very misleading. Suppose the total population is 1,000,000. Then of that population 1000 have the nadgers but the rest do not. Suppose they are all tested. The test will give the wrong answer 1% of the time.

So of the 1000 people that have the nadgers, it will be positive in 99% of them, that is 990 cases. On the other hand consider the 999,000 people that do not have the nadgers. It will (erroneously) prove positive for 1% of them, that is 9,990 cases (these are known as False Positives). So altogether it proves positive in 10,980 cases out of which only 990 actually have the nadgers. So the final outcome is that if the test proves positive then there is less than about a 1 in 10 chance of having the disease. You need a second opinion!

However, if the test proves negative then you can be fairly happy. Of the 1000 people with the nadgers it will prove negative in only 10 cases. And of the 999,000 people that do not have the nadgers it will be negative in 989,010 cases. So if the test proves negative the chance of actually having the nadgers is only about 1 in 100,000.

These figures are shown in the table below with the false positives highlighted in red.

So medical testing for conditions can be very misleading because of false positives unless there is a reason to suppose that the condition is actually present or the test is extremely reliable. And that is why arbitrary general screening for many conditions is just not worthwhile.

The sums come out very differently if there are some other grounds for suspecting that you have the nadgers. For example, it might be that a symptom

test	healthy	sick	total
+ve	**9990**	990	10980
−ve	989010	10	989020
total	999000	1000	1000000

Testing for the Nadgers.

test	healthy	sick	total
+ve	30	990	1020
−ve	2970	10	2980
total	3000	1000	4000

Testing for the Nadgers with green ears.

of the nadgers is that your ears go green. Indeed it might be that everyone with the nadgers has green ears but not vice versa. Suppose that 25% of people with green ears have the nadgers.

Now suppose we apply the test to a population of 4000 people with green ears. Of these 1000 will have the nadgers. Again the test will prove positive for 990 of them. And of the 3000 that do not have the nadgers it will (incorrectly) prove positive for 30 of them. So of the 1020 cases showing positive the nadgers will be present in 990.

So if you have green ears and the test is positive then it is 30 to 1 certain that you do indeed have the nadgers. Better get it cured. See the table above.

Which box has the prize?

SOME YEARS AGO a TV show had the following puzzle. We are shown three boxes which we can denote as A, B, and C. We are told that one box contains a fantastic prize, a golden crown (or perhaps a chocolate bar). The other boxes contain nothing.

We are asked to guess which box has the prize. We make a guess (perhaps box A). The presenter (Mr M) then opens one of the other boxes (perhaps B) and Lo and Behold it is empty. We are asked do we want to change our selection?

The question is should we or not? Does it really matter? We know that the prize is not in B but it seems as if it might equally be in A or C so there is no point in changing. We could sharpen the problem by saying that to change one's mind costs a small fee (somewhat less than the prize of course).

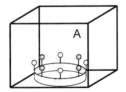 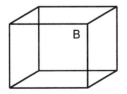 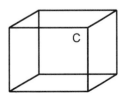

Which box has the prize?

If we conclude that it is worth changing our mind and that the prize is worth £300, then how much would it be worth paying to change our mind?

Suppose the prize is in box A. Then there are two situations to consider. Either we choose A or we don't. Consider these separately.

Choosing A. This will arise with probability 1/3. Mr M can then show us B or C. If we change our mind and select the other box then we lose. If we don't change our mind, we win.

Choosing B or C. This will arise with probability 2/3. Mr M then has no choice, if we choose B then he must show us C and vice versa. If we change our mind and select box A then we win. If we don't change our mind then we lose.

So in one-third of the cases, we lose if we change our mind, but in two-thirds of the cases we win if we change our mind. So we should change our mind.

How much should we pay to change our mind? If we don't change our mind then we expect to win £300 with probability 1/3 and £0 with probability 2/3. So that scenario is worth £100. If we do change our mind then we expect to win £300 with probability 2/3 and £0 with probability 1/3. So that scenario is worth £200. Therefore it is worth the difference between these namely £200–£100 = £100 to be allowed to change our mind. Although we might not like the risk of losing £100.

Of course if we don't know what the prize is worth then it is more difficult. But it seems that we should pay 1/3 of what we think the prize is worth to be allowed to change our mind.

A nasty variation on this puzzle is as follows. There are two boxes as below. We are told that one box has an amount of money in it and the other box has twice that amount of money. Perhaps box A has £1 and box B has £2. But we aren't told these amounts.

We are asked to open a box – we do so and find £2. Now we don't know whether the other box has £1 or £4. We are then asked do we want to change our mind? We can keep the £2 or open the other box and take what is in it. It looks like it is evens whether the other box has £1 or £4. If we change boxes then we might get £1 or £4 with equal probability. Hence by changing boxes we get £2.50 on average. So it is worth changing boxes. That must be ridiculous. Where is the flaw?

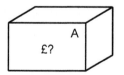
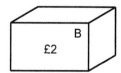

Which box has the most?

Let's look at this more slowly. There are two boxes, A has £1 and B has £2. Fact.

If we choose A, we see the £1 and decide to change our mind and so get £2. This happens with probability 1/2.

If we choose B, we see the £2 and decide to change our mind and so get £1. This happens with probability 1/2.

So with probability 1/2 we get £2 and with probability 1/2 we get £1. On average we get £1.50.

Now suppose we don't change our mind. Again on average we get £1.50. So changing our mind makes no difference.

So what was the flaw in the earlier argument? If we open a box and find £2 then we argue that the other box has £1 or £4. By changing our mind we think we either gain £2 or lose £1. Suppose the actual sums are £1 and £2. If we open the £1 box we think we might gain £1 or lose 50p. What actually happens is that if we open the box with £2 then although we don't realise it, by changing our mind we take the lose £1 option, and if we open the box with £1 then by changing our mind we take the win £1 option. So by changing our mind we either win £1 or lose £1 so it doesn't matter. Another way of looking at this is to note that if we open the bigger box by chance then we have more to lose by changing or mind.

Hmmm. Part of the trouble with this problem is that we are not really dealing with probabilities at all. Box A and box B have in them what they have. We are not forecasting a probabilistic event like tossing a coin which is in the future but simply making an observation of the current situation. This is statistics and the proper term to use is likelihood rather than probability.

Craps and poker

THIS LECTURE finishes with a few words about two typical gambling games, namely craps and poker. Craps is played with dice and poker with cards. Craps is very popular in the US but almost unknown in the UK. Craps is basically a pure game of luck much as Snakes and Ladders.

The essence of craps is that one player S (for shooter) rolls two dice and the other players bet on the outcome. Rolling is done in two phases. On the first roll

> If it is 2, 3, or 12 then S loses and the game is over. S is said to have crapped out!
>
> If it is 7 or 11 then S wins and the game is over.
>
> If it is some other number then this number is known as the Point and S continues to roll until either the Point is rolled again which is a win for S or a 7 is rolled in which case S loses.

Surprisingly, it is almost evens as to whether S wins or loses. The calculation is somewhat laborious. Remember that the probability of either 2 or 12 is 1/36 each, of 3 or 11 is 2/36, of 4 or 10 is 3/36, of 5 or 9 is 4/36, of 6 or 8 is 5/36 and of 7 is 6/36.

So the probability of an initial crap is $1/36 + 2/36 + 1/36 = 1/9$.

The probability of 7 or 11 is $6/36 + 2/36 = 2/9$, twice as likely as a crap.

The probability of something else and so entering the Point phase is 2/3.

In the Point phase, we keep rolling until we get a 7 or the Point. The probability of getting a 7 first depends upon what the Point is. If it is 4 then getting a 7 first is more likely than if the Point is 5 or 6. Consider the possibilities in turn.

If the Point is 4 then on each roll the probability of getting a 4 is 3/36 whereas the probability of getting 7 is 6/36. If it is neither, we roll again. So the probability of getting a 7 first is 6/9. The same applies if the Point is 10.

If the Point is 5 then on each roll the probability of getting a 5 is 4/36 whereas the probability of getting 7 is 6/36. If it is neither, we roll again. So the probability of getting a 7 first is 6/10. The same applies if the Point is 9.

If the Point is 6 then on each roll the probability of getting a 6 is 5/36 whereas the probability of getting 7 is 6/36. If it is neither, we roll again. So the probability of getting a 7 first is 6/11. The same applies if the Point is 8.

To find the overall probability of getting a 7 we have to consider the probability of each Point number and multiply by the probability of getting a 7 in that case.

In deciding the probability of a particular Point we have to remember that only 2/3 of the initial rolls lead to a Point. So they have to be scaled so that the sum of all the point probabilities is one. The probability of getting either 4 or 10 is 9/36; for 5 or 9 it is 12/36; for 6 or 8 it is 15/36.

We find that the overall probability of the Point phase ending on a 7 is

$9/36 \times 6/9 + 12/36 \times 6/10 + 15/36 \times 6/11$

$= 9/54 + 12/60 + 15/66 = 294/495$

And of course the probability of the Point phase not ending on a 7 is 201/495.

So altogether we have a lose if we have an initial crap (1/9) or a 7 turns up in the Point phase ($294/495 \times 2/3$). These add to

$1/9 + 588/1485 = 753/1485$ or about 50.7% lose

In contrast we have a win if we get an initial 7 or 11 (2/9) or the Point phase does not end on a 7 ($201/495 \times 2/3$). These add to

$2/9 + 402/1485 = 732/1485$ or about 49.3% win

It is finely balanced. Maybe years of experimentation took place to get such a result out of a curious sequence. In practice the game is complex because one is allowed to bet either way and also to bet once the Point phase has been entered.

We will now briefly look at the game of poker. All players are dealt five cards. In the simplest game (Straight Poker) they then bet on those hands. In the more usual Draw Poker, they then take it in turn to change as many cards as they like just once. We start with a quick reminder of the possible combinations in increasing order of value.

Five cards	A hand where the cards are unrelated. Such hands themselves are ordered with Queen High beating Jack High etc.
One pair	Two cards are the same. Two Queens beats Two Jacks. If two hands both have the same pairs then the highest of the other three cards is used to decide which hand is higher.
Two pairs	Two Kings and Two Threes beats Two Queens and Two Tens. If two hands both have the same pairs then the fifth card is used to decide the higher hand.
Threes	Such as Three Jacks which beats Three Tens.
Straight	This is five cards of adjacent values irrespective of suit. Examples are 3, 4, 5, 6, 7 which is beaten by 9, 10, J, Q, K. An ace can be high or low so A, 2, 3, 4, 5 or 10, J, Q, K, A are possible Straights.
Flush	All five cards of the same suit. In comparing two Flushes, the highest card in each are compared.
Full house	Three of one kind and two of another such as Three Queens and Two Fives which beats Two Queens and Three Fives.
Fours	Four of a kind such as Four Aces.
Straight Flush	Five cards of adjacent values of the same suit. This is a combination of a Straight and a Flush.
Royal Flush	Ace, King, Queen, Jack, Ten of the same suit.

The probabilities of being dealt these hands are shown in the table overleaf. The total number of combinations is given by the formula for ^nC_r explained at the beginning of this lecture and so is 52! / (47! × 5!) = 52.51.50.49.48/120 which indeed is 2,598,960.

It is surprising that the odds against most hands are so long. Straight Poker is thus rather boring since in an evening of several hands it is quite likely that nothing more exciting than a few Pairs and maybe Threes will be encountered.

But remember that Poker is really all about betting and bluffing. The players in turn can drop out, see, or raise. Raise means to increase the stake. See means to maintain the stake. When no player wishes to raise further or some agreed limit is reached, then the player with the winning combination takes all the staked money.

Knowing when and how often to bluff would take us into the realms of game theory which is somewhat tricky.

Hand	Odds against	Combinations
Nothing	about evens	1,302,540
One pair	1¼ to 1	1,098,240
Two pairs	20 to 1	123,552
Threes	46 to 1	54,912
Straight	254 to 1	10,200
Flush	508 to 1	5,108
Full house	693 to 1	3,744
Fours	4,164 to 1	624
Straight flush	72,192 to 1	36
Royal flush	649,739 to 1	4
Total		2,598,960

The odds of being dealt various hands in Poker.

In Draw Poker we have to consider the various ways of improving a hand. It is a complex story so we shall just summarize the more obvious situations according to the initial deal.

The most likely situation is that we are dealt nothing at all. But there are bound to be two of the same suit so just exchange the other three. The possible outcomes are shown in the table below.

So there is a reasonable chance of getting a pair. But the most likely outcome is that the hand will not be improved at all. The chance of a flush (which we might have hoped for by keeping the two cards of the same suit) is somewhat remote.

If the initial hand has one pair then again exchange three cards. The possible outcomes are shown in the table opposite. Again the most likely outcome is that

Hand	Odds against	Combinations
Nothing	about evens	8,198
One pair	1½ to 1	6,744
Two pairs	22 to 1	711
Threes	42 to 1	377
Flush	97 to 1	165
Full house	900 to 1	18
Fours	8107 to 1	2
Total		16,215

The probabilities of improving two of the same suit by changing three cards.

Hand	Odds against	Combinations
Nothing better	2½ to 1 on	11,559
Two pairs	5 to 1	2,592
Threes	8 to 1	1,854
Full house	97 to 1	165
Fours	359 to 1	45
Total		16,215

The probabilities of improving One Pair by changing three cards.

the hand will not be improved. But there is a moderate hope that we might end up with Two Pairs or Threes; it's slightly better than 3 to 1 against. Anything better is remote. Beware that it might be wise not to discard an Ace since there is a chance of the hand becoming worse off.

If we have four players and they play 10 games in an evening then there are 40 hands to consider. Two Pairs and Threes are almost bound to occur. And a Flush or Full House are not that unlikely.

Here is a brief summary of what to discard according to the initial deal.

One Pair
: Generally discard three cards. If one is an Ace consider retaining the Ace and discard two cards. However, if four cards including one of the Pair form the basis of a Flush or Straight, consider sacrificing the Pair since there is a 1 in 4 chance of getting another Pair and a not insignificant chance of a Straight or Flush.

Two pairs
: Discard the odd card. A 1 in 11 chance of Full house.

Threes
: Discard two cards. A 1 in 23 chance of Fours and 1 in 15 of Full house.

Full house
: Do nothing.

Fours
: Could discard the odd card especially if it is low.

Flush, Straight
: Do nothing.

Otherwise
: If four or three cards form a basis for a Straight or Flush then discard one or two cards. If the same number of cards form the basis for both then go for a Flush. Finally, keep the two cards of the same suit and discard three cards or withdraw from the hand.

Good luck!

Further reading

A CLASSIC WORK on probabilities is *An Introduction to Probability Theory and its Applications* by Feller. A somewhat lighter book which considers various games is *The Mathematics of Games and Gambling* by Packel.

See also Appendix B for further properties of Pascal's triangle and related geometrical topics and Appendix C for a discussion on stochastic processes which should be consulted before attempting Exercise 3 below.

Exercises

1   Suppose that you are visiting Jail whilst playing Monopoly (London version). On leaving you move to a location dictated by the roll of two dice. The next twelve locations are

   1 Pall Mall
   2 Electric Company
   3 Whitehall
   4 Northumberland Avenue
   5 Marylebone Station
   6 Bow Street
   7 Community Chest
   8 Marlborough Street
   9 Vine Street
   10 Free Parking (a corner)
   11 Strand
   12 Chance

   What is the probability of landing on one of Bow Street, Marlborough Street, or Vine Street? Calculate the probability of landing on them individually and then add these to get the answer.

2   In a class of N students what is the probability of two or more of them having the same birthday? Give a general formula. Evaluate for $N = 10$, 20, and 30 as a percentage. Assume 365 days in a year and that none of them was born on February 29 in a leap year.

3* (Read Appendix C before attempting this exercise.) In the game of Risk, two groups of armies fight for territories. At each turn the attacker and the defender roll dice. The attacker can roll up to 3 dice, the defender up to 2 dice. The attacker must have at least one more army than the number of dice he rolls. The defender must have at least as many armies as the number of dice he rolls. For example if the attacker has 3 armies then he can only roll 1 or 2 dice; if the defender has 6 armies then he can roll 1 or 2 dice. We will assume that both attacker and defender roll as many dice as possible on each turn.

On each turn, the defender wins in the case of a draw. If both play several dice the highest are compared and then the second. For example playing the maximum dice (6,5,4) v (5,5) results in 6 beating 5 (attacker wins) and 5 drawing with 5 (defender wins), so the net result is both lose an army. Similarly (5,4,4) v (5,4) results in the attacker losing two armies. And (6,5) v (5,4) results in the defender losing two armies.

They keep playing until the attacker kills all the defender's armies or the attacker gets fed up (typically has only one army left and so cannot continue). The general question is, if the attacker has m armies and the defender has n armies what is the probability of the attacker killing all the defender's armies? In particular, what is the probability of the attacker winning if

1) the attacker starts with 2 armies and the defender has 1,

2) the attacker starts with 3 armies and the defender has 2,

3) the attacker starts with 4 armies and the defender has 2.

4 Fractions

FRACTIONS come in various forms. There are vulgar (or simple) fractions, decimal fractions, and continued fractions to mention but a few. In this lecture we look at vulgar fractions and how they were used by the Egyptians and then at decimal fractions and, in particular, recurring decimal fractions. We also look at square and cube roots and then at continued fractions. We start, however, by summarizing the kinds of real numbers.

Real numbers

BY REAL NUMBERS we mean as opposed to complex numbers which will be the subject of a later lecture. Moreover, we will consider only positive real numbers and also exclude zero for the moment. So the numbers we are considering are those that might naturally occur in measurements. There are four categories

Whole numbers or integers, such as 3,

Rational numbers, such as 4¼,

Irrational numbers, such as $\sqrt{2}$,

Transcendental numbers, such as π.

These all turn up quite naturally in simple primitive measurements.

Whole numbers occur in simple counting such as would occur when selling 3 cows to your neighbour.

Rational numbers occur when dividing whole entities, such as dividing 17 loaves among 4 soldiers. This gives each soldier 4¼ loaves.

Irrational numbers occur in certain measurements such as finding the diagonal of a square whose side is 1 cubit. The diagonal is $\sqrt{2}$ cubits.

Transcendental numbers occur in other measurements such as finding the length of the circumference of a circle whose diameter is 1 unit. The length of the circumference is π units.

The key point about an irrational number is that it cannot be expressed as the ratio of two integers, or in other words as a rational fraction a/b. It is said that a pupil of the Pythagorean school first proved that $\sqrt{2}$ could not be equal to any rational fraction a/b where a and b are whole numbers. This upset the Pythagoreans who believed that everything could be expressed as the ratio of integers and it is said that the poor pupil was put to death by drowning. The proof goes something like this.

Suppose $\sqrt{2} = a/b$ and that we have cancelled out any common factors. This means that

$$a^2 = 2b^2$$

The right hand side is therefore even and so the left hand side must also be even. This means that a cannot be odd and so must be even. And since b and a have no common factors, b must be odd. Since a is even we can write it as $a = 2c$ and so

$$4c^2 = 2b^2$$

Now we cancel out 2 and get

$$2c^2 = b^2$$

And now we can deduce that b is even which is a contradiction because we just proved that b is odd.

So we are forced to conclude that $\sqrt{2}$ cannot be written as the ratio of two whole numbers. This process of proof by a contradiction was also encountered in the lecture on Amicable Numbers and is known as *Reductio ad Absurdum*.

Irrational numbers come in two categories: algebraic numbers and transcendental numbers. Numbers that are the solutions of equations such as

$$3x^3 - 4x^2 + x + 7 = 0$$

are algebraic numbers. So $\sqrt{2}$ is an algebraic number since it is the root of the equation

$$x^2 - 1 = 0$$

Some numbers such as π are not the root of any such equation and these numbers are called transcendental numbers.

So there are really just two categories, rational and irrational numbers, and each has two subcategories. The rational numbers are either whole numbers or not, and the irrational numbers are either algebraic or transcendental.

The rest of this lecture will look at some properties of rational numbers.

Vulgar fractions

FRACTIONS can be written in various forms. Thus we can write one half as 0.5 or as ½. The form 0.5 is called a decimal fraction whereas ½ is called a vulgar fraction. Vulgar is used here in the sense of common or ordinary.

By definition any rational number can be written as a vulgar fraction. Sometimes the term proper fraction is used to denote a fraction less than 1 such as $3/7$ whereas the term improper fraction is used to denote a fraction such as $9/7$ which is normally written as $1 2/7$.

We normally cancel out common factors, so we don't write $3/9$ but $1/3$. This can make life confusing. Traditional imperial measurements for tools might go $1/32$, $1/16$, $3/32$, $1/8$, $5/32$, $3/16$, $7/32$, $1/4$ and so on. These are of course in increments of $1/32$. But the many changes of denominator obscures this as was mentioned in the lecture on Measures.

Note that we use the terms numerator and denominator for the numbers above and below the line respectively. Numerator comes from the Latin, numare, to count and indicates the number of parts. Denominator comes from the Latin nominare, to name and indicates the name or value of the parts.

Egyptian fractions

VARIOUS PAPYRI have come down to us concerning mathematics. Perhaps the most important is the Rhind papyrus written in about BC 1700 by a scribe called Ahmes; it is in the British Museum. Others are the Moscow papyrus and the Berlin papyrus both written around BC 1900.

A fascinating introduction to the Rhind papyrus will be found in volume one of *The World of Mathematics* by James Newman. This is a collection of essays on various mathematical topics. The article on the papyrus is actually a reprint of an article in Scientific American in 1952. It shows an extract from the papyrus and has a worked example giving the original so-called hieratic script (like handwriting), the corresponding hieroglyphics, and a modern transliteration.

The Egyptians used base 10 although they did not have a place system and did not use zero. But we can use our normal notation to explain some of their techniques. The hieroglyphic form will be explained in the lecture on Notations.

Multiplication was done by repeated doubling. So to multiply 15 by 13 we write

$$1 \times 15 = 15$$
$$2 \times 15 = 30$$
$$4 \times 15 = 60$$
$$8 \times 15 = 120$$

and then noting that $13 = 8 + 4 + 1$ we just select those rows and add the corresponding results

$$1 \times 15 = 15$$
$$4 \times 15 = 60$$
$$\underline{8 \times 15 = 120}$$
$$13 \times 15 = 195$$

In order to see that $13 = 8 + 4 + 1$ we can write it in binary as 1101. A good way to do this is to keep dividing 13 by 2 and noting the remainders (this topic is discussed in more detail in the lecture on Notations).

$13 \div 2 = 6$ rem 1; $6 \div 2 = 3$ rem 0; $3 \div 2 = 1$ rem 1; $1 \div 2 = 0$ rem 1.

The remainders form the binary digits starting at the unit end. This can be coupled with the repeated doubling. The rows to be excluded from the addition are those with even quotients such as 6.

$1 \times 15 = 15$ 13
$2 \times 15 = 30$ 6   rem 1
$4 \times 15 = 60$ 3   rem 0
$8 \times 15 = 120$ 1   rem 1

So we cross out the row with 6 and add up the others.

Division was done in a similar way. Suppose we want to divide 143 by 13. We build a table of doublings of 13 as follows

$1 \times 13 = 13$
$2 \times 13 = 26$
$4 \times 13 = 52$
$8 \times 13 = 104$

Now we note that $143 = 104 + 26 + 13$. So we just select those rows and again add the corresponding results.

$1 \times 13 = 13$
$2 \times 13 = 26$
$\underline{8 \times 13 = 104}$
$11 \times 13 = 143$

So the answer is 11. But if the division is not exact then we have to use fractions as well.

The Egyptians had a very peculiar approach to fractions. Apart from $2/3$ which was treated specially, they worked entirely in terms of reciprocals. Thus instead of $3/4$, they would use the fact that $3/4 = 1/2 + 1/4$.

They didn't write it like this but simply represented a reciprocal as the denominator in their script with a dot over it. In transcriptions a line over is frequently used so we can write $\overline{2}\,\overline{4}$ as shorthand for $1/2 + 1/4$.

Now any fraction can be decomposed into this form by simply taking away the largest reciprocal possible at any stage. Let's do that with 2/5, 3/5, and 4/5.

2/5 is bigger than 1/3 but less than 1/2 so we take away 1/3. Now $2/5 - 1/3 = 6/15 - 5/15 = 1/15$. So we are done, $2/5 = 1/3 + 1/15$.

4 Fractions   73

3/5 is greater than 1/2, so we take that away. Now 3/5–1/2 = 6/10–5/10 = 1/10. So we are done, 3/5 = 1/2+1/10.

4/5 is greater than 1/2, so we take that away. Now 4/5–1/2 = 8/10–5/10 = 3/10. So now we need to work out 3/10. Now 3/10 is greater than 1/4 but less than 1/3, so we take away 1/4. Now 3/10–1/4 = 6/20–5/20 = 1/20. So we are done and finally we have, 4/5 = 1/2+1/4+1/20.

Or in other words

$2/5 = \bar{3}\ \overline{15}$

$3/5 = \bar{2}\ \overline{10}$

$4/5 = \bar{2}\ \bar{4}\ \overline{20}$

It was often the case that for fractions greater than 2/3, this was used as the first term. So since 4/5 is greater than 2/3, we take that away. Now 4/5–2/3 = 12/15–10/15 = 2/15. Now we work out 2/15. This is greater than 1/8 but less than 1/7, so we take away 1/8. Now 2/15–1/8 = 16/120–15/120 = 1/120. So we have

$4/5 = \bar{\bar{3}}\ \bar{8}\ \overline{120}$

where 2/3 is represented as 3 with a double bar over it.

Dividing loaves

THE RHIND PAPYRUS contains many sample problems. Many are about dividing loaves among several men. Problem 6 concerns the subdivision of 9 loaves among 10 men.

They each need 9/10 of a loaf. If we were faced with this problem, then we might cut one tenth off each loaf, give the big pieces to nine of the men and then give all the small pieces to the tenth man. But this would probably cause a riot since it would not be clear that they were being treated fairly.

A better approach might be to halve five of the loaves and give each man half a loaf to start with. Then divide the remaining loaves into thirds giving twelve pieces – give ten to the ten men leaving two pieces of a third of a loaf over. Cut each of these into five and give one piece to each man. Each man then gets three pieces being a half, a third, and a fifteenth of a loaf respectively as shown diagrammatically overleaf.

So we have

$1/2 + 1/3 + 1/15 = 9/10 = \bar{2} + \bar{3} + \overline{15}$

This is clearly fair. Each man has exactly the same number of pieces of bread of the same size – they don't have to know the sizes of the pieces to see that they have been treated fairly. Moreover, we see the value of the unit fraction approach to doing the subdivision.

74 Nice Numbers

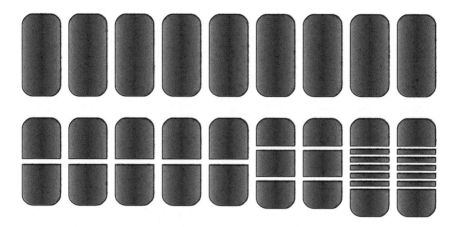

Dividing nine loaves into ten portions – our solution.
Each portion is a half plus a third plus a fifteenth of a loaf.

Curiously, this is not the answer in the Rhind papyrus which is in fact to give each man two-thirds of a loaf, a fifth of a loaf, and then a thirtieth of a loaf. This corresponds to the subdivision

$$2/3 + 1/5 + 1/30 = 9/10 = \bar{\bar{3}} + \bar{5} + \overline{30}$$

This seems a bit awkward since there are not enough loaves to give each man a two-third piece. Instead, we can start by dividing seven loaves into thirds and give each man two of these. The remaining third can be cut into ten pieces and the other two loaves are divided into five pieces each as shown below.

This doesn't seem nearly so neat since altogether there are 40 pieces of bread whereas the previous solution involved only 30. Crumbs!

Moreover, it is clear that there is still a practical problem with the crusts. How does one cut the middle third so that it is equitable with the end pieces? No doubt tricky and would give the person in charge much food for thought!

Dividing nine loaves into ten portions according to the Rhind papyrus.
Each portion comprises two thirds plus a fifth plus a thirtieth of a loaf.

4 Fractions

Curiously, the papyrus doesn't actually explain how the solution is obtained but simply shows that the result is correct by multiplying it by 10 and checking that it correctly gives 9. Thus by doubling we have

```
  1      3̄ 5̄ 3̄0̄
/ 2    1 3̿ 1̄0̄ 3̄0̄
  4      3̿ 2̄ 1̄0̄
/ 8        7 5
```

The normal multiplication process is used and only the lines marked 2 and 8 need to be added together. Let us attempt to struggle through the process of multiplying each line by 2.

It is not at all clear how they multiplied the first line which is 2/3+1/5+1/30 by 2 to get the second line which is 1+2/3+1/10+1/30. Presumably they multiplied each term by 2 and then rearranged as necessary.

Multiplying fractions by 2 was a common requirement and the Rhind papyrus contains a table giving values for all fractions of the form 2/5, 2/7, 2/9, ..., 2/101 (this table is discussed in the next section). For this example we need

$$2/5 = 1/3 + 1/15$$
$$2/15 = 1/10 + 1/30$$

So now we can double 2/3 to get 1+1/3. We double 1/5 to get 2/5 and then use the table. To double 1/30 is easy because 30 is even and so the answer is simply 1/15.

If we gather these all together we get

$$1 + 1/3 + 1/3 + 1/15 + 1/15$$

Now we are allowed to add the two instances of 1/3 to give 2/3, but for the two instances of 1/15 we have to use the double fraction table again and look up 2/15. We end up with

$$1 + 2/3 + 1/10 + 1/30$$

and indeed that is the line against 2 in the table.

Let's try doubling once more. It seems easier this time since 1/10 becomes 1/5 and 1/30 becomes 1/15 and this results in

$$3 + 1/3 + 1/5 + 1/15$$

Sadly, that is not the same as the line against 4 in the table which is

$$3 + 1/2 + 1/10$$

although it does have the same value. Maybe the scribe had certain simplifying rules as well. Maybe he used the table of tenths which is also given in the Rhind

2/10 = 1/5	6/10 = 1/2 + 1/10
3/10 = 1/5 + 1/10	7/10 = 2/3 + 1/30
4/10 = 1/3 + 1/15	8/10 = 2/3 + 1/10 + 1/30
5/10 = 1/2	9/10 = 2/3 + 1/5 + 1/30

The table of tenths.

papyrus as shown above. It shows that 1/3 + 1/15 is 4/10, so adding 1/5 gives 6/10 which is clearly 1/2 + 1/10. Maybe the scribe had a sort of scratchpad for these minor calculations.

Finally, we can double this line and indeed that immediately gives the line against 8 which is

$$7 + 1/5$$

Now we have to add the lines against 2 and 8. This gives

$$7 + 1/5 + 1 + 2/3 + 1/10 + 1/30$$

Consulting the table of tenths again shows that

$$9/10 = 2/3 + 1/5 + 1/30$$

and this exactly matches three of the fractions and so we have

$$7 + 1 + 9/10 + 1/10 = 9$$

and we are done at last. That is all very hard work. Maybe we will never know how the scribes really did their calculations.

Note that the decomposition of a fraction is not unique, for example, we can write 7/29 as either of

$$7/29 = 1/6 + 1/24 + 1/58 + 1/87 + 1/232$$
$$7/29 = 1/5 + 1/29 + 1/145$$

In choosing between different possibilities it appears that the scribes probably followed the following rules

Smaller numbers are preferred, and numbers above 1000 avoided.

An equality with a smaller number of terms is preferred.

No terms can be the same, and they are always written in descending order.

The smallness of the first number is the main consideration but a larger first number is acceptable if that greatly reduces the last number.

Even numbers are preferred to odd numbers although they might be larger and the number of terms might be increased.

4 Fractions

It is hard to choose between the two versions of 7/29. The first version has more terms, and both the first and last terms are larger which all count against. On the other hand it has only one odd term whereas all terms of the second version are odd. Even numbers are presumably preferred because it makes doubling trivial and as we have seen doubling is an important part of multiplication.

It has been conjectured that perhaps the Egyptians knew about the Fibonacci numbers and their use in approximating the golden number. Recall from Lecture 2 on Amicable Numbers that the Fibonacci numbers are

1, 1, 2, 3, 5, 8, 13, 21, 34, 55, 89, 144, 233, ...

where each number is the sum of the previous two.

The ratios of pairs are closer and closer approximations to the golden number $\tau = (\sqrt{5} + 1)/2 = 1.6180...$. Note carefully that the inverse of τ is simply $\tau - 1$. So we get

1/2 = 0.5
2/3 = 0.666...
3/5 = 0.6
5/8 = 0.625
8/13 = 0.615...
13/21 = 0.619...
21/34 = 0.617...
34/55 = 0.618...

and so on. The values are alternately above and below the inverse of the golden number. We can write the ratios as Egyptian fractions thus

1/2 = 1/2
2/3 = 1/2 + 1/6
3/5 = 1/2 + 1/10
5/8 = 1/2 + 1/10 + 1/40
8/13 = 1/2 + 1/10 + 1/65
13/21 = 1/2 + 1/10 + 1/65 + 1/273
21/34 = 1/2 + 1/10 + 1/65 + 1/442
34/55 = 1/2 + 1/10 + 1/65 + 1/442 + 1/1870
55/89 = 1/2 + 1/10 + 1/65 + 1/442 + 1/3026
89/144 = 1/2 + 1/10 + 1/65 + 1/442 + 1/3026 + 1/12816

This is quite fascinating. Items alternately add a new term or change the last term. When a new term is added, the denominator is obtained by multiplying the two adjacent numbers of the series. Thus for 13/21 we add 273 = 13×21 and for 34/55 we add 1870 = 34×55. When a term is changed, it becomes the product of two alternate terms in the series. Thus for 21/34 we change 273 into 442 = 13×34 and for 55/89 we change 1870 into 3026 = 34×89.

Perhaps a nicer way of doing this is to permit subtraction of terms rather than changing a term and we then get

1/2 = 1/2
2/3 = 1/2 + 1/6
3/5 = 1/2 + 1/6 − 1/15
5/8 = 1/2 + 1/6 − 1/15 + 1/40
8/13 = 1/2 + 1/6 − 1/15 + 1/40 − 1/104
13/21 = 1/2 + 1/6 − 1/15 + 1/40 − 1/104 + 1/273

and so on. We now find that the term added or subtracted is simply the product of two adjacent terms of the Fibonacci series in each case.

The table of $2/n$

THE TABLE OPPOSITE gives the decompositions of $2/n$ from the Rhind papyrus. The first and last columns give n for convenience. The column headed final result gives the value as unit fractions but for simplicity the bar above each number is omitted. Thus in the case of 31 we have

$2/31 = 1/20 + 1/124 + 1/155$

The second and any subsequent terms are always closely related. In this case 124 is 4×31 and 155 is 5×31 and this is captured in the column marked interim which shows this factorization.

It is not really known how the table was constructed. Clearly we start with an inspired guess of the first term, in this case 20. So to compute the remaining terms we need to evaluate 2/31 minus 1/20, perhaps thus

$$\frac{2}{31} - \frac{1}{20} = \frac{40-31}{31 \times 20} = \frac{9}{31 \times 20} = \frac{4}{31 \times 20} + \frac{5}{31 \times 20} = \frac{1}{31 \times 5} + \frac{1}{31 \times 4}$$

which gives the terms 1/124 and 1/155.

Some entries in the table are surprising. For any value $2/(2n-1)$ we have

$$\frac{2}{2n-1} = \frac{1}{n} + \frac{1}{n(2n-1)}$$

and so putting $n = 3$ and $n = 8$, we have

2/5 = 1/3 + 1/15   as in the table
2/15 = 1/8 + 1/120   but table has 1/10 + 1/30

So that general formula gives the value in the table for 2/5 but 2/15 is different. Maybe they disliked 1/8 + 1/120 because of the size of the last term.

4 Fractions 79

n	decompose	interim	final result	n
5	= 3+(1+1/2)+1/2	3+(5×3)	= 3 + 15	5
7	= 4+2+1	4+(7×4)	= 4 + 28	7
9	= 6+3	6+(9×2)	= 6 + 18	9
11	= 6+4+1	6+(11×6)	= 6 + 66	11
13	= 8+4+1	8+13(4+8)	= 8 + 52 + 104	13
15	= 10+5	10+(15×2)	= 10 + 30	15
17	= 12+4+1	12+17(3+4)	= 12 + 51 + 68	17
19	= 12+6+1	12+19(4+6)	= 12 + 76 + 114	19
21	= 14+7	14+(21×2)	= 14 + 42	21
23	= 12+8+3	12+(23×12)	= 12 + 276	23
25	= 15+10	15+(25×3)	= 15 + 75	25
27	= 18+9	18+(27×2)	= 18 + 54	27
29	= 24+4+1	24+29(2+6+8)	= 24 + 58 + 174 + 232	29
31	= 20+10+1	20+31(4+5)	= 20 + 124 + 155	31
33	= 22+11	22+(33×2)	= 22 + 66	33
35	= 30+5		= 30 + 42	35
37	= 24+12+1	24+37(3+8)	= 24 + 111 + 296	37
39	= 26+13	26+(39×2)	= 26 + 78	39
41	= 24+16+1	24+41(6+8)	= 24 + 246 + 328	41
43	= 42+1	42+43(2+3+7)	= 42 + 86 + 129 + 301	43
45	= 30+15	30+(45×2)	= 30 + 90	45
47	= 30+15+2	30+47(3+10)	= 30 + 141 + 470	47
49	= 28+14+7	28+(49×4)	= 28 + 196	49
51	= 34+17	34+(51×2)	= 34 + 102	51
53	= 30+20+3	30+53(6+15)	= 30 + 318 + 795	53
55	= 30+20+5	30+(55×6)	= 30 + 330	55
57	= 38+19	38+(57×2)	= 38 + 114	57
59	= 36+18+3+2	36+59(4+9)	= 36 + 236 + 531	59
61	= 40+20+1	40+61(4+8+10)	= 40 + 244 + 488 + 610	61
63	= 42+21	42+(63×2)	= 42 + 126	63
65	= 39+26	39+(65×3)	= 39 + 195	65
67	= 40+20+5+2	40+67(5+8)	= 40 + 335 + 536	67
69	= 46+23	46+(69×2)	= 46 + 138	69
71	= 40+20+10+1	40+71(8+10)	= 40 + 568 + 710	71
73	= 60+10+3	60+73(3+4+5)	= 60 + 219 + 292 + 365	73
75	= 50+25	50+(75×2)	= 50 + 150	75
77	= 44+22+11	44+(77×4)	= 44 + 308	77
79	= 60+15+4	60+79(3+4+10)	= 60 + 237 + 316 + 790	79
81	= 54+27	54+(81×2)	= 54 + 162	81
83	= 60+20+3	60+83(4+5+6)	= 60 + 332 + 415 + 498	83
85	= 51+34	51+(85×3)	= 51 + 255	85
87	= 58+29	58+(87×2)	= 58 + 174	87
89	= 60+20+6+3	60+89(4+6+10)	= 60 + 356 + 534 + 890	89
91	= 70+14+7		= 70 + 130	91
93	= 62+31	62+(93×2)	= 62 + 186	93
95	= 60+30+5	60+95(4+6)	= 60 + 380 + 570	95
97	= 56+28+7+4+2	56+97(7+8)	= 56 + 679 + 776	97
99	= 66+33	66+(99×2)	= 66 + 198	99
101	= 101+0	101+101(2+3+6)	= 101 + 202 + 303 + 606	101

Other typical entries in the 2 times table are

2/19 = 1/12 + 1/76 + 1/114
2/59 = 1/36 + 1/236 + 1/531
2/97 = 1/56 + 1/679 + 1/776

However, the formula would give these three entries as

2/19 = 1/10 + 1/190
2/59 = 1/30 + 1/1770
2/97 = 1/49 + 1/4753

in which the second term has rather large values for the denominator. So the table was presumably crafted to give acceptable results according to the criteria mentioned earlier.

A detailed discussion will be found in the fascinating book *Mathematics in the Time of the Pharaohs* by Richard Gillings. He mentions that an analysis searching for all possible decompositions into unit fractions was performed on the KDF9 in the University of Sydney in 1967. (No doubt in KDF9 Algol – I remember it well). In the case of 31, there were 164 decompositions of which one had just two terms namely 1/16 + 1/496. Possibly this was rejected because 496 is too large. Of those with three terms, (20, 124, 155) has the smallest terms and was presumably chosen for this reason despite the odd term.

Note that if n is divisible by 3 then the decomposition is straightforward. This is because 2/3 = 1/2 + 1/6. The decomposition of 101 is surprising and arises because 2 = 1 + 1/2 + 1/3 + 1/6. Any number can be decomposed in this way but the first term would then not be even which is clearly undesirable.

The method of false position

A NUMBER of the problems in the Rhind papyrus are of the form "A fraction $\bar{p}$ of me is added to me to give N. What am I?"

In other words we have to solve the linear equation

$x + x/p = N$

The approach taken is to make a guess of x, see what $x + x/p$ becomes, compare it with N and adjust x accordingly.

Problem 24 is of this form with p being 7 and N being 19. The technique is to assume the false answer 7 for x; that is we assume that x is actually equal to p. That makes x/p equal to 1 and ensures that we are just dealing with whole numbers at this stage. In this example that would make N equal to 7 + 1 = 8.

But we want N to be 19 so now we have to multiply the false answer 7 by 19 divided by 8.

So we start by dividing 19 by 8 thus

$$\begin{array}{rl} 1 \times 8 = & 8 \\ /\ 2 \times 8 = & 16 \\ \bar{2} \times 8 = & 4 \\ /\ \bar{4} \times 8 = & 2 \\ /\ \bar{8} \times 8 = & 1 \end{array}$$

and noting that 19 = 16+2+1 we find that the answer is

$$2 \quad \bar{4} \quad \bar{8} \quad \text{or} \quad 2 + 1/_4 + 1/_8 = 2^3/_8$$

Now we multiply this by 7 by repeated doubling thus

$$\begin{array}{rl} /\ 1 \times & 2 \quad \bar{4} \quad \bar{8} \\ /\ 2 \times & 4 \quad \bar{2} \quad \bar{4} \\ /\ 4 \times & 9 \quad \bar{2} \end{array}$$

adding the rows and gathering terms together gives

$$\begin{array}{rl} 7 \times & 15 \quad (\bar{2} \quad \bar{2}) \quad (\bar{4} \quad \bar{4}) \quad \bar{8} \\ 7 \times & 16 \quad \bar{2} \quad \bar{8} \end{array}$$

so that the final answer is

$$16 \quad \bar{2} \quad \bar{8} \quad \text{or} \quad 16 + 1/_2 + 1/_8 = 16^5/_8$$

Perhaps we had better check this

$$16^5/_8 \times 1^1/_7 = {}^{133}/_8 \times 8/_7 = {}^{133}/_7 = 19$$

so it works.

Decimal fractions

WE NOW LEAVE the strange world of Egyptian fractions and consider familiar decimal fractions. They can be classified into three kinds

Terminating, such as 0.25; these represent a rational fraction whose denominator has prime factors which are just 2 and 5.

Recurring, such as 0.3636363636...; these represent a rational fraction whose denominator has prime factors other than 2 and 5.

Neither recurring nor terminating; these represent irrational numbers such as $\sqrt{2}$ or π.

```
        142857
    _____
7 ) 1.0000000...
    7
    ─
    30
    28
    ──
    20
    14
    ──
    60
    56
    ──
    40
    35
    ──
    50
    49
    ──
    10
```

Dividing 1.0 by 7 using long division until the remainder repeats.

Recurring fractions are interesting. For example, consider 0.3636.... Let us suppose its value is x, and consider $100x$.

$x = 0.363636...$
$100x = 36.363636...$

Subtract these and the recurring bits cancel, giving

$99x = 36$

and cancelling the common factor of 9, we finally get

$x = 4/11$

It is clear that we can do this with any recurring fraction and so any such is equal to a rational fraction.

Now consider the reverse process. Suppose we wish to calculate 1/7 as a decimal fraction. The long division process goes as shown in the diagram above.

At each stage we get a remainder. There are only 6 possible remainders, namely 1, 2, 3, 4, 5, 6. The first remainder is 3 and after six stages one of the remainders must repeat and the process will then continue indefinitely.

Eventually we get a remainder of 1 and then 3 occurs once more with the result that we have

$1/7 = 0.142857142857...$

The remainders occur in a cycle and in fact all the sevenths consist of the same sequence but starting at different points in the cycle, thus

$4/7 = 0.571428571428...$
$6/7 = 0.857142857142...$

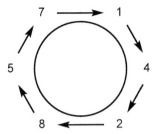

The cycle of digits of sevenths.

If we add 1/7 and 6/7 we get

$1/7 = 0.142857142857...$
$6/7 = 0.857142857142...$
$7/7 = 0.999999999999...$

which is of course 1.

Anyway, it shows that the last 3 digits of the sequence plus the first 3 add to 999. This gives a handy way of remembering the sequence for 1/7. If you remember that π is about $3\frac{1}{7}$ and that $\pi = 3.142$ approximately then you can remember all 6 digits by just subtracting the first 3 from 999.

There is a convenient notation for recurring decimals — we put a dot over the first and last digits of the recurring sequence, so we write

$1/7 = 0.\dot{1}4285\dot{7}$

If we do the same with 1/11 we find that the sequence is not of length 10 but of length 2.

$1/11 = 0.090909... = 0.\dot{0}\dot{9}$

The rule is that the cycle length of the sequence for $1/n$ where n is prime is always a factor of $n-1$. In this case n is 11 so the sequence must be a factor of 10 and 2 is indeed a factor of 10. So in the case of elevenths there are six pairs of fractions thus

$1/11 = 0.\dot{0}\dot{9}$ $10/11 = 0.\dot{9}\dot{0}$
$2/11 = 0.\dot{1}\dot{8}$ $9/11 = 0.\dot{8}\dot{1}$
$3/11 = 0.\dot{2}\dot{7}$ $8/11 = 0.\dot{7}\dot{2}$
$4/11 = 0.\dot{3}\dot{6}$ $7/11 = 0.\dot{6}\dot{3}$
$5/11 = 0.\dot{4}\dot{5}$ $6/11 = 0.\dot{5}\dot{4}$

Another very familiar couple of recurring fractions are those for the thirds

$1/3 = 0.\dot{3}$ $2/3 = 0.\dot{6}$

In this case the sequence length is just 1 and only one dot is required.

84   Nice Numbers

The next one is 1/13. In this case there are two sequences whose length is 6

$1/13 = 0.\dot{0}7692\dot{3}$ $2/13 = 0.\dot{1}5384\dot{6}$

$3/13 = 0.\dot{2}3076\dot{9}$ $5/13 = 0.\dot{3}8461\dot{5}$

$4/13 = 0.\dot{3}0769\dot{2}$ $6/13 = 0.\dot{4}6153\dot{8}$

$9/13 = 0.\dot{6}9230\dot{7}$ $7/13 = 0.\dot{5}3846\dot{1}$

$10/13 = 0.\dot{7}6923\dot{0}$ $8/13 = 0.\dot{6}1538\dot{4}$

$12/13 = 0.\dot{9}2307\dot{6}$ $11/13 = 0.\dot{8}4615\dot{3}$

The next prime number after 13 is 17 and 1/17 has a full cycle of 16 digits thus

$1/17 = 0.\dot{0}588235294117647$

And the expansions of the next few prime fractions, 19, 23, 29 also have full cycles of 18, 22, and 28 digits respectively. But when we come to 31 we find that there are two distinct cycles of 15 digits and when we come to 37 we find that

$1/37 = 0.\dot{0}2\dot{7}$

and perhaps surprisingly has a cycle of just 3 digits. Other fairly short cycles are those for 41 (5 digits), 73 (8 digits) and 79 (13 digits).

This seems all rather random, so just what is the rule for the length of the cycle of a prime fraction?

The length of the cycle is the smallest power of 10 which when divided by the prime number gives remainder 1.

Let's try that on a few numbers. First take 3

$10 \div 3 = 3$ remainder 1

so it is just the first power of 10 and so the sequence length is 1. Correct.
Now try 7

$10 \div 7$ has remainder 3
$100 \div 7$ has remainder 2
$1000 \div 7$ has remainder 6
$10000 \div 7$ has remainder 4

But this is hard work. There is an easier way, we just take the remainder each time and multiply that by 10 and divide again. In other words we do arithmetic modulo 7 as explained in the lecture on Amicable Numbers.
We get

$10^1 = 3$ mod 7
$10^2 = 2$ mod 7   ($3 \times 10 = 30$ and divide by 7 giving 4 remainder 2)

$10^3 = 6 \mod 7$ ($2 \times 10 = 20$ and divide by 7 giving 2 remainder 6)
$10^4 = 4 \mod 7$
$10^5 = 5 \mod 7$
$10^6 = 1 \mod 7$ got it

So the answer is 6 and indeed the sequence length for 1/7 is 6.

And similarly for 11. Since 100 has remainder 1 on dividing by 11 it follows that 1/11 has a cycle of length 2.

The very short sequence for 37 is because 37 is a factor of 999. And so 1000 divided by 37 has remainder 1 and the sequence for 1/37 therefore has length 3 because $1000 = 10^3$. The key of course is that 37 is a factor of 111 and so of 999.

So here is perhaps a revealing way of looking at it. The factors of 99 are 9 and 11. So 1/11 etc. is the only prime fraction with cycle 2. The 9 is just noise and anyway 1/9 has cycle 1.

So let's methodically go through all the 9s. Now $999 = 9 \times 111$ and as we have just seen $111 = 3 \times 37$ so 37 is the only prime with cycle 3.

Then $9999 = 9 \times 1111$ and $1111 = 11 \times 101$ so 101 should have a cycle of 4. And indeed it does

$1/101 = 0.\dot{0}09\dot{9}$

Now consider 99999. The factors of 11111 are 41×271. That should mean that 1/41 and 1/271 have a cycle of 5. Indeed

$1/41 = 0.\dot{0}243\dot{9}$

$1/271 = 0.\dot{0}036\dot{9}$

The next one is 999999. Now 111111 has lots of factors. It is $3 \times 7 \times 11 \times 13 \times 37$. This is the first time we have encountered 7 and 13 and confirms that they have cycle length of 6.

And then consider 9999999. Now 1111111 = 239×4649 which is amazing. So 1/239 and 1/4649 should have cycles of length 7.

And indeed they do

$1/239 = 0.\dot{0}04184\dot{1}$

$1/4649 = 0.\dot{0}00215\dot{1}$

And so it goes on. Prime numbers and factorizations are pretty random so it is to be expected therefore that the cycles are erratic.

It is important to note that we have been dealing with fractions $1/n$ where n is prime. The rule that the sequence length is a factor of $n-1$ does not apply if n is not prime. Good examples are given by 1/81 and 80/81 which are

$1/81 = 0.\dot{0}1234567\dot{9}$

$80/81 = 0.\dot{9}8765432\dot{0}$

1	1	
2	11	11
3	111	3.37
4	1111	11.101
5	11111	41.271
6	111111	3.7.11.13.37
7	1111111	239.4649
8	11111111	11.73.101.137
9	111111111	3^2.37.33367
10	1111111111	11.41.271.9091
11	11111111111	21649.513239
12	111111111111	3.7.11.13.37.101.9901
13	1111111111111	53.79.265371653
14	11111111111111	11.239.4649.909091
15	111111111111111	3.31.37.41.271.2906161
16	1111111111111111	11.17.73.101.137.5882353
17	11111111111111111	2071723.5363222357
18	111111111111111111	3^2.7.11.13.19.37.52579.333667
19	1111111111111111111	1111111111111111111
20	11111111111111111111	11.41.101.271.3541.9091.27961

Factors of 11, 111, 1111 etc.

and have a cycle length of 9. We will encounter the ratio 80/81 which corresponds to a musical interval known as the syntonic comma when we discuss Music in Lecture 9. Another interesting example is 1/27 which is

$$1/27 = 0.0\dot{3}\dot{7}$$

and has a cycle length of 3. Remember that 37 is a favourite number! And moreover that 1/37 is 0.027027.... This is all because 27×37 = 999.

The rule for when *n* is prime is summarized in the table above which gives the factors of the numbers 1, 11, 111 and so on. These are sometimes referred to as repunits. So in order to find the cycle of a prime fraction we simply look down the table until we find its first occurrence and the corresponding number of 1s is the answer.

Thus in the case of 101 it first occurs as a factor of 1111 and so has a 4-digit sequence which is

$$1/101 = 0\dot{0}9\dot{9}$$

Note that some factors keep cropping up. This is because 111 is of course a factor of 111111 and 111111111 and so on.

Roots

AN OPERATION THAT is very like long division is finding a square root. I am sure that it is hardly known these days because one can simply use a pocket calculator. But before looking at it we return to the Egyptians since they clearly did know a bit about square roots.

The Berlin papyrus contains problems which (if understood correctly) involve finding square roots. Gillings mentions one problem as follows

The area of a square of 100 square cubits is equal to the sum of the areas of two smaller squares. The side of one is $\bar{2}\ \bar{4}$ times the side of the other. Find the lengths of the sides of the two unknown squares.

The method used is essentially that of false position mentioned earlier. We assume that the bigger square has side 1; it then follows that the other square has side $\bar{2}\ \bar{4}$. We now multiply $\bar{2}\ \bar{4}$ by itself. This is

$$(\bar{2} + \bar{4}) \times (\bar{2} + \bar{4}) = \bar{4} + \bar{8} + \bar{8} + \overline{16} = \bar{4} + \bar{4} + \overline{16} = \bar{2} + \overline{16}$$

giving $\bar{2}\ \overline{16}$ which is therefore the area of the smaller square. So the area of the two squares is $1\ \bar{2}\ \overline{16}$.

We now have to scale up to make this 10. To find the scale factor we take the square root of 100 (which is 10) and the square root of $1\ \bar{2}\ \overline{16}$ (which is $1\ \bar{4}$ as shown below). And then divide one by the other to give 8. This is the required scale factor. So the bigger square has side 8 obtained by multiplying the assumed 1 by the scale factor. The smaller square then has side given by multiplying $\bar{2}\ \bar{4}$ by 8 which is 6.

This is simply the Pythagorean triangle $10^2 = 8^2 + 6^2$. It seems that the Egyptians knew the theorem a long time before it was named after Pythagoras.

Sadly, the papyrus does not show how the square root of $1\ \bar{2}\ \overline{16}$ was obtained. Nor indeed how any of the calculations were done.

Gillings suggests that they must have had tables of squares. An extract from 6 to 7 might have been as shown overleaf. It is very unbalanced. Attempting to extrapolate to find perhaps the square root of 41 could have driven one mad.

The square of $1\ \bar{4}$ is
$1 + \bar{4} + \bar{4} + \overline{16}$
which is
$1\ \bar{2}\ \overline{16}$.

6	36	6 $\bar{2}$	42 $\bar{4}$
6 $\bar{8}$	37 $\bar{2}$ $\bar{64}$	6 $\bar{\bar{3}}$	44 $\bar{3}$ $\bar{9}$
6 $\bar{6}$	38 $\bar{36}$	6 $\bar{2}$ $\bar{4}$	45 $\bar{2}$ $\bar{16}$
6 $\bar{5}$	38 $\bar{3}$ $\bar{15}$ $\bar{25}$	6 $\bar{2}$ $\bar{3}$	46 $\bar{4}$ $\bar{3}$ $\bar{9}$
6 $\bar{4}$	39 $\bar{16}$	6 $\bar{\bar{3}}$ $\bar{4}$	48 9 16
6 $\bar{3}$	40 $\bar{9}$	7	49

Some Egyptian squares from 6 to 7.

Let us try nevertheless. It is obviously somewhere between 6 $\bar{3}$ and 6 $\bar{2}$. We could try 6 $\bar{3}$ $\bar{10}$. Its square is 41 $\bar{5}$ $\bar{9}$ $\bar{15}$ $\bar{100}$. In our world that is 41.38... which is too large. Now try 6 $\bar{3}$ $\bar{15}$. Its square is 40 $\bar{\bar{3}}$ $\bar{9}$ $\bar{10}$ $\bar{15}$ $\bar{90}$ $\bar{225}$ which (apart from being really nasty) is 40.96 and so is too small but we are getting near. Now try 6 $\bar{3}$ $\bar{14}$ (its square is 41.02...) so we can settle on that. Its square in unit fractions is – we leave that to the patient reader!

Returning to modern times we consider how the square root can be calculated by hand. Let us find the square root of 76825225.

First divide the number up into pairs of digits starting at the right hand end to get 76'82'52'25 and consider the first two (maybe just one) digits – in this case 76. Now 8 squared is 64 and 9 squared is 81 so the answer must lie between 8000 and 9000. In other words the first digit is an 8. The theory of the method relies upon the rule that $(a+b)^2 = a^2 + 2ab + b^2 = a^2 + b(2a+b)$.

So if a is 8000 we need to find b so that $(2a+b) \times b$ is 76825225 minus 8000 squared. We do this by stages and the work is usually laid out as shown below. Put the 8 in the answer line, square it and take it away from the first pair of digits and draw the next pair down. This gives 1282. Now we double what we know so far to give 16 (that is $2a$ in the formula) and then need to find the next digit, shown in the first stage as a ?, such that multiplying 16? by ? can be taken away from 1282. (This is the situation on the left.) The next digit is 7 since 167×7 is

```
        8                    8 7                      8 7 6 5
   ─────────             ─────────              ─────────
   8)76 82 52 25         8)76 82 52 25          8)76 82 52 25
     64                    64                     64
     ──                    ──                     ──
 16?)12 82             167)12 82              167)12 82
                           11 69                  11 69
                           ─────                  ─────
                      174?)1 13 52           1746)1 13 52
                                                  1 04 76
                                                  ───────
                                             17525) 8 76 25
                                                    8 76 25
                                                    ───────
                                                         0
```

Three stages in finding the square root of 76825225.

```
                                        8  7  6  5
                                    ─────────────────
                                    673 373 097 125
                                    512
                                    ───
  3×80           240                161 373
   ×87         20880
   +7²            49
               ─────
               20929         ×7     146 503
                                    ───────
  3×870         2610                14 870 097
   ×876      2286360
    +6²           36
             ───────
             2286396         ×6     13 718 376
                                    ──────────
 3×8760        26280                1 151 721 125
  ×8765     230344200
     5²            25
           ─────────
           230344225         ×5     1 151 721 125
                                    ─────────────
                                                 0
```

Finding the cube root of 673373097125.

1169 but 168×8 is 1344 and 1344 is too much. So we now put the 7 in the answer line subtract 1169 from 1282 and draw the next two digits down.

We now have an answer so far of 87 and a new dividend of 11352. Double 87 to give 174 and find a digit ? so that multiplying 174? by ? can be taken away from 11352. This is the situation in the second diagram. The answer is 6 so the next digit is 6 and so on. The final answer is 8765.

There is a similar but much more tedious technique for finding cube roots. It depends upon the formula $(a+b)^3 = a^3 + 3a^2b + 3ab^2 + b^3 = a^3 + b(3a^2+3ab+b^2)$.

We will try it on 673373097125. The first thing to do is to mark the number off into groups of three digits thus 673'373'097'125. The first three digits are 673. Now 8 cubed is 512 and 9 cubed is 729 so the answer must lie between 8000 and 9000. In other words the first digit is 8. See the diagram above.

So if a is 8000 we seek b such that $(3a^2+3ab+b^2)\times b$ is the original number minus 8000 cubed. We can write this as $(3a(a+b)+b^2)\times b$ which helps a little.

We take 512 from 673 and then draw the next three digits down. This results in 161373. Let us guess the next digit is also 8. So we have a = 80 and try b = 8. The term $3a$ is then 240 and $(a+b)$ is 88. We get (240×88 + 64)×8 = (21120+64)×8 = 21184×8 = 169472. This is bigger than 161373 so 8 is too much. Let's try 7. This gives (240×87 + 49)×7 = (20880+49)×7 = 20929×7 = 146503. That is less than 161373 so 7 is the next digit. This gives remainder of 14870 and we then draw the next three digits down to give 14870097.

The process then repeats with $a = 870$. An initial guess for the next digit can be obtained by dividing 3×870^2 into 14870097. Now 3×870^2 is 2270700 and this goes 6 times but not 7 into 14870097. So trying 6 we get $(2610 \times 876 + 36) \times 6 = (2286360+36) \times 6 = 13718376$. So 6 it is and the remainder is now 1151721 and then we draw the last three digits down to get 1151721125.

Do it once more with $a = 8760$. Initial guess for the next digit is obtained by dividing 3×8760^2 into 1151721125. Now 3×8760^2 is 230212800 and this goes 5 times but not 6 into 1151721125. So trying 5 we get $(26280 \times 8765 + 25) \times 5 = (230344200+25) \times 5 = 115721125$. Gosh that is exactly it. So the answer is 8765.

Continued fractions

CONTINUED FRACTIONS are a curious topic. They figured highly in mathematics examinations in the early nineteenth century then fell out of favour but are now considered rather important. They are very useful for finding convenient approximations and we will use them in the lecture on Music.

Consider a fraction less than 1 such as 7/31. We can write this as

$$\frac{1}{31/7} = \frac{1}{4^3/_7} = \cfrac{1}{4 + \cfrac{1}{7/3}} = \cfrac{1}{4 + \cfrac{1}{2^1/_3}} = \cfrac{1}{4 + \cfrac{1}{2 + \cfrac{1}{3}}}$$

This is usually written in one of various shorthand notations such as

$$\cfrac{1}{4 +} \cfrac{1}{2 +} \cfrac{1}{3}$$

or even more succinctly as [4, 2, 3]. If a number is greater than 1 such as $5^7/_{31}$ then we can write it as [5; 4, 2, 3] with a semicolon after the integer part.

All fractions of the form a/b terminate as continued fractions whereas many do not if expressed as decimal fractions. Clearly $1/10$ is simply [10], the other tenths are as follows

2/10 = [5] 3/10 = [3, 3] 4/10 = [2, 2] 5/10 = [2]
6/10 = [1, 1, 2] 7/10 = [1, 2, 3] 8/10 = [1, 4] 9/10 = [1, 9]

which reveals a rather curious mess. Although $1/10$ terminates as a decimal fraction we saw above that $1/7$ is recurring. But as a continued fraction it is simply [7]. The other sevenths are

2/7 = [3, 2] 3/7 = [2, 3] 4/7 = [1, 1, 3] 5/7 = [1, 2, 2] 6/7 = [1, 6]

Note that $(n-1)/n$ is always $[1, n-1]$. We have mentioned that using traditional fractions such as $1/4$, $5/16$, $3/8$, and so on for measurements can be confusing since their order is not immediately obvious. But just look at the sequence from $1/32$ to $31/32$ as continued fractions

$1/32 = [32]$	$1/16 = [16]$	$3/32 = [10, 1, 2]$	$1/8 = [8]$
$5/32 = [6, 2, 2]$	$3/16 = [5, 3]$	$7/32 = [4, 1, 1, 3]$	$1/4 = [4]$
$9/32 = [3, 1, 1, 4]$	$5/16 = [3, 5]$	$11/32 = [2, 1, 10]$	$3/8 = [2, 1, 2]$
$13/32 = [2, 2, 6]$	$7/16 = [2, 3, 2]$	$15/32 = [2, 7, 2]$	$1/2 = [2]$
$17/32 = [1, 1, 7, 2]$	$9/16 = [1, 1, 3, 2]$	$19/32 = [1, 1, 2, 6]$	$5/8 = [1, 1, 1, 2]$
$21/32 = [1, 1, 1, 10]$	$11/16 = [1, 2, 5]$	$23/32 = [1, 2, 1, 1, 4]$	$3/4 = [1, 3]$
$25/32 = [1, 3, 1, 1, 3]$	$13/16 = [1, 4, 3]$	$27/32 = [1, 5, 2, 3]$	$7/8 = [1, 7]$
$29/32 = [1, 9, 1, 2]$	$15/16 = [1, 15]$	$31/32 = [1, 31]$	$1 = [1]$

That really is an even more confusing incoherent mess! Beauty in mathematics is largely about discovering patterns. To find good examples of beauty in continued fractions we consider square roots.

The continued fractions for square roots do not terminate but have recurring patterns; for example $\sqrt{7} = [2; 1, 1, 1, 4, 1, 1, 1, 4, 1, 1, 1, 4, ...]$. For convenience the first and last of the pattern can be marked with a dot above as with recurring decimal fractions. The values for the first few are

$\sqrt{2} = [1; \dot{2}]$	$\sqrt{3} = [1; \dot{1}, \dot{2}]$	$\sqrt{5} = [2; \dot{4}]$
$\sqrt{6} = [2; \dot{2}, \dot{4}]$	$\sqrt{7} = [2; \dot{1}, 1, 1, \dot{4}]$	$\sqrt{8} = [2; \dot{1}, \dot{4}]$
$\sqrt{10} = [3; \dot{6}]$	$\sqrt{11} = [3; \dot{3}, \dot{6}]$	$\sqrt{12} = [3; \dot{2}, \dot{6}]$
$\sqrt{13} = [3; \dot{1}, 1, 1, 1, \dot{6}]$	$\sqrt{14} = [3; \dot{1}, 2, 1, \dot{6}]$	$\sqrt{15} = [3; \dot{1}, \dot{6}]$
$\sqrt{17} = [4; \dot{8}]$	$\sqrt{18} = [4; \dot{4}, \dot{8}]$	$\sqrt{19} = [4; \dot{2}, 1, 3, 1, 2, \dot{8}]$
$\sqrt{20} = [4; \dot{2}, \dot{8}]$	$\sqrt{21} = [4; \dot{1}, 1, 2, 1, 1, \dot{8}]$	$\sqrt{22} = [4; \dot{1}, 2, 4, 2, 1, \dot{8}]$
$\sqrt{23} = [4; \dot{1}, 3, 1, \dot{8}]$	$\sqrt{24} = [4; \dot{1}, \dot{8}]$	$\sqrt{26} = [5; \dot{10}]$
$\sqrt{27} = [5; \dot{5}, \dot{10}]$	$\sqrt{28} = [5; \dot{3}, 2, 3, \dot{10}]$	$\sqrt{29} = [5; \dot{2}, 1, 1, 2, \dot{10}]$

Although it might not be obvious they all have the same overall structure. After the initial integer, the pattern repeats indefinitely. Moreover, the last number of the pattern is always twice the initial integer and the rest of the pattern is palindromic. Thus in the case of 14, the initial integer is 3 and the last number of the recurring pattern is 6. The rest of the pattern is 1, 2, 1 which is palindromic.

Note that roots near an exact square follow a standard structure, thus

$$\sqrt{(n^2+1)} = [n; \dot{2n}] \qquad \sqrt{(n^2+2)} = [n; \dot{n}, \dot{2n}]$$
$$\sqrt{(n^2-1)} = [n-1; \dot{1}, \dot{2n-2}] \qquad \sqrt{(n^2-2)} = [n-1; \dot{1}, n-2, 1, \dot{2n-2}]$$

So square roots become recurring continued fractions of a certain style. We can ask what is the value of recurring continued fractions in general.

As a simple example consider

$$x = [1, 2, 1, 2, ...]$$

If we write this as

$$x = \cfrac{1}{1+} \cfrac{1}{2+} \cfrac{1}{1+} ...$$

it should be clear that the bit following the first 2 is also x so we have

$$x = \cfrac{1}{1+} \cfrac{1}{2+x} = \frac{2+x}{3+x} \quad \text{and rearranging we get}$$

$$x^2 + 2x - 2 = 0$$

and then using the standard formula for the quadratic equation $ax^2 + bx + c = 0$ which we might remember is

$$x = [-b \pm \sqrt{(b^2 - 4ac)}] / 2a$$

we find that $x = -1 + \sqrt{3}$, which agrees with the value given earlier for just $\sqrt{3}$. As another example we find that $[1, 3, 1, 3, ...]$ is $(\sqrt{21} - 3) / 2$. We can use the same approach for any recurring period. Consider

$$x = [1, 2, 3, 1, 2, 3, ...] \quad \text{so that}$$

$$x = \cfrac{1}{1+} \cfrac{1}{2+} \cfrac{1}{3+x}$$

and unwinding this we finally get $x = (7 + 2x) / (10 + 3x)$ giving

$$3x^2 + 8x - 7 = 0 \quad \text{and finally} \quad x = (\sqrt{42} - 4) / 3$$

It should be clear that we always get a quadratic equation for x no matter how long the recurring period. As a consequence a cube or other higher roots, logarithms and so on cannot possibly have a recurring continued fraction. They are just a bit of a mess. Here are some examples

π = [3; 7, 15, 1, 292, 1, 1, ...]
$\sqrt[3]{2}$ = [1; 3, 1, 5, 1, 1, 4, 1, ...]
log 10 = [2; 3, 3, 3, 1, 1, 3, 6, ...]
$\log_2(3/2)$ = [0; 1, 1, 2, 2, 3, 1, 5, ...]
$2^{1/12}$ = [1; 16, 1, 4, 2, 7, 2, 13, ...]

An important property of continued fractions is that they can be truncated and then give good approximations. Indeed, the approximations are the best for the size of the denominator. Such truncated values are called the convergents and alternate convergents are above and below the original number. For example, considering the square root of 2 we have

$[1; 2] = 3/2 = 1.5$
$[1; 2, 2] = 7/5 = 1.4$
$[1, 2, 2, 2] = 17/12 = 1.41666...$
$[1, 2, 2, 2, 2] = 41/29 = 1.41379...$
$[1, 2, 2, 2, 2, 2] = 99/70 = 1.41428...$

whereas the correct value is 1.41421.... As another example, consider π

$[3; 7] = 22/7 = 3.1428571...$
$[3, 7, 15] = 333/106 = 3.1415094...$
$[3, 7, 15, 1] = 355/113 = 3.1415929...$

whereas the correct value is 3.14159265....

There is a simple rule for calculating one convergent from the previous two. Suppose the continued fraction is $[a_0; a_1, a_2, a_3, ...]$ and that the convergents are $c_0 = [a_0;] = n_0/d_0$, $c_1 = [a_0; a_1] = n_1/d_1$, $c_2 = [a_0; a_1, a_2] = n_2/d_2$, and so on. (Note that we are using n for the numerators and d for the denominators.) The rule is simply

$$n_i = n_{i-1} \times a_i + n_{i-2} \quad \text{and} \quad d_i = d_{i-1} \times a_i + d_{i-2}$$

In other words the numerator of the next convergent is obtained by multiplying the numerator of the previous convergent by the next term of the continued fraction and adding the numerator of the convergent before that. The same mechanism applies to the denominator.

In the case of π the zero convergent, c_0, is 3/1, and c_1 is 22/7. The next term is 15 and so

$$n_2 = 22 \times 15 + 3 = 333 \quad \text{and} \quad d_2 = 7 \times 15 + 1 = 106$$

and so the next convergent is 333/106 as stated above. And similarly

$$n_3 = 333 \times 1 + 22 = 355 \quad \text{and} \quad d_3 = 106 \times 1 + 7 = 113$$

giving 355/113.

An important point is that a convergent is a particularly good approximation if the first term ignored is large. Thus 22/7 is a good approximation to π because the term ignored is 15 and 355/113 is amazingly good because the term ignored is 292. We will use such approximations in the lecture on Music.

Two continued fractions deserve a special mention. One is that for e, the base of natural logarithms. It is

$$e = [2; 1, 2, 1, 1, 4, 1, 1, 1, 8, 1, 1, 1, 1, 16, ...]$$

and although this does not recur in the traditional sense it does have a strange repetitive pattern.

The other, the simplest of all, is that for the golden number $\tau = (1 + \sqrt{5})/2$ which was mentioned earlier when discussing Egyptian fractions and is

$$\tau = [1; 1, 1, 1, 1, ...]$$

The convergents for τ are

$[1; 1] = 3/2$
$[1; 1, 1] = 5/3$
$[1; 1, 1, 1] = 8/5$
$[1, 1, 1, 1, 1] = 13/8$

which are the ratios of pairs of Fibonacci numbers.

Beware that continued fractions are quite unlike recurring decimal fractions although we are using a similar notation. Thus in the case of decimal fractions we have

$0.\dot{1} = 1/9$ $0.\dot{2} = 2/9$ $0.\dot{3} = 3/9$ $0.\dot{6} = 6/9$

and these values are clearly related in that they are all multiples of 1/9. But in the case of continued fractions we have

$[\dot{1}] = (\sqrt{5}-1)/2$ $[\dot{2}] = \sqrt{2}-1$ $[\dot{3}] = (\sqrt{13}-3)/2$ $[\dot{6}] = \sqrt{10}-3$

and there is no obvious relationship between the values at all.

The eye of Horus

WE CONCLUDE this lecture with a few gentle remarks about Egyptian weights and measures. There is a legend concerning the gods. Horus was the son of Osiris who was killed by his brother Seth. In revenge Horus killed his uncle but in the fight lost an eye. The broken parts of the eye were later restored by the god Thoth. Note that Isis was the mother of Horus and the wife (and sister, tut-tut) of Osiris.

The various parts of the eye were used to denote fractions as shown opposite. Note that they add to 63/64; the missing 1/64 is said to represent the blood lost. Curiously, it seems that these Horus-eye fractions were only used to denote a fraction of a hekat which was a measure of grain, about half a peck or a gallon.

The unit of length was the cubit of about 18 inches. The hayt was 100 cubits. The arura was an area of 10,000 square cubits, about 2,500 square yards and so roughly half an acre.

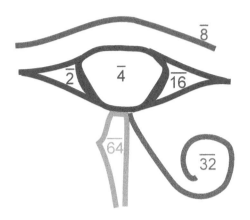

The eye of Horus showing the fractions.

The word hekat also turns up in Problem 79 in the Rhind Papyrus where it seems to have a quite different meaning. This says that a man has seven houses. Each house has seven cats. Each cat has seven mice. Each mouse has seven spelts (ears of corn). Each spelt has seven hekats (grains). What is the total?

Man	1
Houses	7
Cats	49
Mice	343
Spelts	2401
Hekats	16807
Total	19608

An alternative way to do this which was known to the Egyptians is to compute the sequence of partial sums. Thus to add together $1 + 3 + 9 + 27 + 81$ we can do the sums of the partial series by first computing $1 + 3$, then $1 + 3 + 9$, then $1 + 3 + 9 + 27$, and so on. The key point is that

$$1 + 3 + 9 + 27 = ((1 + 3 + 9) \times 3) + 1$$

So having got the sum so far, we multiply by the factor 3 and then add 1. Indeed we do not actually have to multiply 27 by 3 to get 81.

In the case of the seven cats and so on we can work as follows

$1 + 7 = 8$

$8 \times 7 + 1 = 56 + 1 = 57$

$57 \times 7 + 1 = 399 + 1 = 400$

$400 \times 7 + 1 = 2800 + 1 = 2801$

$2801 \times 7 + 1 = 19607 + 1 = 19608$

which gives the same answer.

It has been conjectured that this puzzle is the origin of

> As I was going to St Ives,
> I met a man with seven wives,
> Each wife had seven cats,
> Each cat had seven kits,
> ...
> How many were going to St Ives?

This is a trick question because the answer is one, just the speaker, the others were coming away from St Ives!

Further reading

THE TOPIC of Egyptian mathematics as a whole is discussed at length in *Mathematics in the Time of the Pharaohs* by Richard Gillings. A good introduction and in particular the Rhind papyrus are discussed in volume 1 of *The World of Mathematics* by James Newman. There are also articles in Historia Mathematica. One is *Calculating the daily bread: Rations in theory and practice* by Annette Imhausen (in vol 30 no 1); another is *Were the Fibonacci Series and the Golden Section Known in Ancient Egypt?* by Corinna Rossi and Christopher Tout (in vol 29 no 2).

A useful book is *The Penguin Dictionary of Curious and Interesting Numbers* by David Wells. It contains tables of decimal fractions of the primes and the factors of the repunits up to R_{40} (that is a series of 40 ones).

Continued fractions are discussed in detail in *Elementary Number Theory* by Kenneth Rosen; this includes a table of square roots up to 100. The golden number and its various applications are discussed in *Gems of Geometry* by the author.

Exercises

1. Multiply 19 by 21 the Egyptian way.
2. Find Egyptian fraction representations of 2/7, 3/7, 4/7, 5/7 and 6/7. Write your answers using bars.
3. Find the recurring decimal representation of 1/73 and confirm that it has a cycle of length 8.

5 Time

THIS LECTURE concerns both times and dates. Knowing about the seasons is important to the development of an agricultural economy (rather than just being hunter/gatherers). And as we have become industrialized, we have found the need for great precision in measuring time.

Basic rhythms

PRIMITIVE MAN will have been aware of three basic rhythms, the day, the month, and the year as follows.

The day – with alternating periods of light and dark caused by the rotation of the Earth on its axis.

The month – defined by the rotation of the Moon around the Earth and noticeable in that it waxed and waned and controlled the tides.

The year – defined by the rotation of the Earth around the Sun and the cause of the variation in climate and the amount of daylight and hence the seasons.

As we know there are about 365¼ days in the year. As a consequence most calendar years have 365 days but some years are leap years with 366 days in order to get the correct average. A more accurate value is 365.2422.

There are about 29½ days in a month as measured from New Moon to New Moon. Dividing 365.25 by 29.5 gives 12.38.... That's a bit irritating. But anyway man has settled for 12 months in most cultures and the Moon just has to put up with not being quite in synchronization.

And more lately we have divided the day into 24 equal hours and so on although early man only counted time during daylight hours.

In astronomical terms these numbers are all wrong. It looks like the Earth goes around on its axis in 24 hours because we measure the day basically by noon to noon. But since the Earth goes around the Sun in a year there is a day different roughly averaged out over the year. In fact the Earth goes around on its axis about 366¼ times a year. And each rotation on its axis takes about 23hours 56mins 4secs – this is known as the sidereal day. The day as measured by 24hrs is known as the solar day. Incidentally, the terms sidereal and solar come from the Latin words sidus, sideris, star, and sol, solis, the sun.

Moreover, the Earth doesn't really go around the Sun in 365.2422 solar days because of the "precession of the equinoxes". The seasons are caused by the fact that the axis of the Earth is tilted by some 23.5° to the plane of its orbit around the Sun. However, this axis rotates much like the axis of a spinning top rotates.

98 Nice Numbers

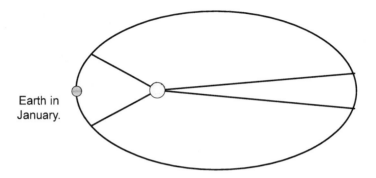

Kepler's third law says equal areas are swept in equal times.

It takes about 25,800 years to do one rotation so we don't feel sick. So the seasons slowly shift with respect to the fixed stars and complete a cycle once every 25,800 years. The consequence is that the sidereal year is actually $1\frac{1}{25,800}$ times a solar year and this works out at 365.2564 solar days.

Basically our day is defined by noon to noon or midnight to midnight. However, the Earth's orbit around the Sun is not a perfect circle but a gentle ellipse. Sometimes the Earth is nearer the Sun and sometimes it is further away. When nearer to the Sun (in January) the Earth travels faster in its orbit. One of Kepler's laws states that equal areas are swept out in equal times. As a consequence the extra sidereal day in a year is not distributed equally throughout the year. Thus although noon to noon is 24 hours on average, sometimes it is less and sometimes it is more. The variation due to this effect is about 7½ minutes either way. Incidentally, the diagram above greatly exaggerates the ellipticity.

Another cause of variation is due to the inclination (or tilt) of the Earth's axis. This inclination is the cause of the seasons. The longest day in terms of daylight is called the summer solstice and the shortest day is the winter solstice. The word solstice derives from the Latin sol, the sun, plus the past participle, status, of sistere, to stand still. So they are the days when the sun stands still, in other words is neither getting higher nor lower in the sky. The summer solstice is around June 20 and the winter solstice is around December 21. The days which have exactly 12 hours of daylight are called the equinoxes (again from the Latin meaning equal nights). They are the vernal equinox around March 20 and the autumnal equinox around September 22. Vernal is from the Latin vernus, of the spring. Note that these dates vary from year to year partly because of leap years; those given are for 2016.

However, as well as causing the seasons, the tilt also causes a variation of about 10 minutes either way in the length of the day from noon to noon as measured on a sundial and is zero at the equinoxes and solstices. These two variations are shown on the diagram opposite. That due to the elliptic orbit is shown in red and that due to the tilt is shown in blue; the combined effect is shown in black.

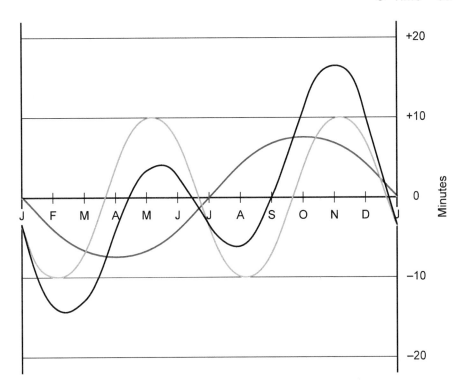

The equation of time.

The result is that a sundial is only correct at certain times of the year – typically on April 15, June 13, September 1, and December 25 but it varies from year to year. The biggest differences are −14 minutes in February and +16 minutes in November. Posh sundials have the corrections marked on them in the form of a graph. The correction is known as "the equation of time". The shortest apparent solar day is about December 21 which is 30 seconds short of 24 hours from noon to noon. We normally talk of the shortest day as being that with the least number of daylight hours. That is also around December 21 (it is a coincidence). There is a period of long days in September when the days are 24 hours plus 20 seconds, and another period in late March/early April when the days are 24 hours plus 18 seconds. June has short days of around 24 hours minus 13 seconds. So it is all a bit wobbly.

Although the shortest day is about December 21, the earliest sunset is on about December 13 and the latest dawn is on about January 2. This shift is due to the equation of time which changes rapidly in December because the Earth is then close to the Sun. That's why the post Christmas mornings seem so bleak. There is a corresponding asymmetry around midsummer but is not so large. Moreover, we are not usually up at dawn to notice the effect!

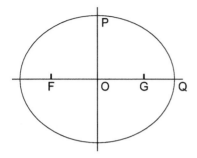

An ellipse with $a = 5$ and $b = 4$.
The eccentricity $e = 0.6$, so $ae = 3$.
The two foci F and G are at the points $(-ae, 0)$ and $(+ae, 0)$.
The point P at $(0, 4)$ is at one end of the minor axis and Q at $(5, 0)$ is at one end of the major axis.

It should also be noted that because the Earth is nearer to the Sun in January, the range of climate in the northern hemisphere is less than that in the southern hemisphere. In order to explain and quantify the effect, we look at the basic properties of an ellipse in general.

The reader might remember that the equation of an ellipse is

$$x^2/a^2 + y^2/b^2 = 1$$

where the lengths of the two axes are $2a$ and $2b$ as in the diagram above. In the case of a circle the values of a and b are the same. The eccentricity e is given by $e^2 = 1 - b^2/a^2$. The two foci are at the points $(\pm ae, 0)$. An ellipse can be drawn by taking a loop of string around the two foci so that a pencil held taut in the loop traces the ellipse.

In the ellipse above, a and b are 5 and 4 respectively and so b^2/a^2 is then $^{16}/_{25}$ which is 0.64. Thus $e^2 = 1 - 0.64 = 0.36$ and so e equals 0.6. The foci are then at the points $(-3, 0)$ and $(+3, 0)$. The point P is at the end of the minor axis so OP is 4 and OF is 3. It follows by Pythagoras that PF is 5. So a string around FGP has total length 16. Note that is consistent with the point Q at the end of the major axis when the string would loop from F to Q and back which is also of length 16.

It is in fact a general property of an ellipse that the distance from a focus to the end of a minor axis is always the same as from the centre O to an end of the major axis, that is $FP = OQ$.

In the case of the orbit of the Earth around the Sun, the eccentricity e is only about 0.0167 and b/a is about 0.99986. The orbit is in fact only a tiny bit non-circular. The distance a is about 149.6 million kms (around 93 million miles). The Sun is at a focus at ae from the centre which is about 2.5 million kms.

So the average distance of the Sun from the Earth is about 149.6 million kms and varies from about 147.1 to 152.1 million kms at the two ends of the major axis – a range of about 3%. However, because of the square law, the range of energy received from the Sun varies by about 6%. Thus it is extra hot in Australia in January not only because it is summer but also because the Sun is nearer – the effects add up whereas in Europe they subtract (there may be other reasons as well such as the fact that the Pacific ocean is huge). However, we need to remember the precession of the equinoxes. At the moment the Earth is nearest to the Sun in January but it changes. In 13,000 years time it will be nearest in July.

Moreover, the eccentricity varies over a period of about 100,000 years and can be as high as 0.07 – the effect would then be much more pronounced.

We stated above that the Moon goes around the Earth in about 29.5 days. That's not quite right either. That is the time from New Moon to New Moon. The topic is complicated partly because the Earth goes around the Sun but also because the plane of the orbit of the Moon around the Earth is not the same plane as the orbit of the Earth around the Sun – there is a tilt of some 5°. These factors mean that the actual sidereal period of the Moon is about 27 days and 7 hours.

We are familiar with the idea of a leap year inserted from time to time to keep the calendar in step with the seasons. Recently, leap seconds have been introduced to compensate for the slowing down of the Earth's rotation. Originally, the second was defined as a certain fraction of the sidereal day. As the Earth slows down, the second actually becomes very slightly longer but until recently we had nothing to really compare it against and so the effect was not observable.

But with the advent of being able to measure atomic frequencies, more accurate standards became possible. Since the 13th General Conference on Weights and Measures in October 1967, the second has been defined as

> The duration of 9,192,631,770 periods of the radiation corresponding to the transition between the two hyperfine levels of the ground state of the caesium-133 atom.

As a consequence, the variations in the length of a day are now more noticeable. The correction is made by adding a leap second every now and then. In effect, the first second after midnight on the day concerned is repeated. This last happened on 30 June 2015. Normally the second after 23:59:59 is 00:00:00 but when a leap second occurs, 23:59:60 or 24:00:00 is inserted between them.

There has recently been talk of abolishing leap seconds and just letting the clocks drift until they are an hour out, when we can then have a worthwhile adjustment of an hour rather like the daylight saving adjustments with which we are quite familiar. Such an adjustment would be many centuries into the future!

The Roman calendar

WE START by considering the origins of the calendar of 12 months now universally used for civil matters. It comes from that devised by the Romans.

The year originally was considered to start around spring when it was time to think about planting seed. The first ten months were named as follows.

1. The first month was Martius and named after Mars, the god of agriculture and war. Mars was also said to be the father of Romulus and thus of all Romans. So a natural choice for the name of the first month.

2   The second month was Aprilis meaning to open up, lay bare – maybe something to do with fertility or the soil. (Incidentally, it has recently been discovered that children conceived in April are the brightest.)

3   And then Maius, named after Maia, the daughter of Atlas, who bore Mercury to Jupiter. In Greek mythology she bore Hermes (= Mercury) to Zeus (= Jupiter). Maia was also goddess of the spring in Greek mythology.

4   The fourth month was Iunius. This was named after the goddess Juno (Greek = Hera), daughter of Saturn, sister *and* wife of Jupiter (tut-tut), guardian deity of all women.

At this point they gave up on the gods and simply used numbers thus

5   Quinctilis, simply the fifth month.
6   Sextilis, the sixth month.
7   September, the seventh month.
8   October, the eight month.
9   November, the ninth month.
10 December, the tenth month.

Originally they gave up there (they weren't counting in the dark days awaiting spring) but later added

11 Ianuarius. This comes from the Latin word Ianus meaning a door or entrance. Janus was the god of gateways and his head had two faces, one on the front and one on the back so he could look both ways. So in terms of the calendar it is the month in which we look forward and back – the gateway to the year.

12 Februarius. This comes from the Latin word Februum which was a festival of cleansing. A hint perhaps to start spring cleaning early.

There is some doubt as to the original length of these months. It is said that they were alternately 29 and 30 days which only gives 354 days in a year. To correct this an extra month of 22 or 23 days was added every other year. But they forgot sometimes so that by 46 BC, Julius Caesar found that it was a complete mess. To put things right, 46 BC had 445 days and was known as the year of confusion. Caesar introduced the Julian calendar in 45 BC and this served well for over a thousand years. (Actually, the situation before the introduction of the Julian calendar was very complex indeed and only roughly as outlined above.)

The Julian calendar introduced leap years with an extra day every fourth year. We add it at the end of February, reflecting the fact that March was the old beginning of the year so if you need to add an extra day, what better place than to add it to the last month of the year. For some reason the Romans did not add the extra day right at the end of February but had February 24 twice (gosh two birthdays for some lucky people). It was known as dies bissextus. However, after

Caesar's death they were confused and made every third year a leap year. This was corrected some thirty years later by omitting the next three leap years. Maybe the error arose because of the Roman habit of counting inclusively.

The original plan it seems was to have the months alternately 30 and 31 days long which would give 366 days. The extra day was chopped off February when not a leap year. The plan was perhaps

31 days	Januarius, Martius, Maius, Quinctilis, September, November
30 days	Februarius, Aprilis, Iunius, Sextilis, October, December

And then Julius Caesar decided that he fancied a month named after him so Quinctilis became Julius. When Augustus became emperor he decided that he wanted a month named after him as well and so Sextilis became Augustus. However, it is said that Augustus didn't fancy just a 30-day month, so he made August have 31 days and stole the day from poor February, thus leaving it with just 28 or 29 rather than 29 or 30. And then to avoid having three 31-day months in succession, September to December were reorganized to give the situation today. We should be grateful that no other emperor fancied a month otherwise February might have been depleted even further.

The numbering of the days of the month in the Roman calendar is curious. Each month has three key named days, the Kalends (the 1st of the month), the Nones and the Ides. In March, May, July, and October, the Nones was the 7th whereas it was the 5th for the other months. The Ides was always 8 days after the Nones. Thus the Ides of March is the 15th. The days between the Kalends, Nones, and Ides are reckoned as so many days before the next key day. Thus the 14th of March was Pridie Idus Martias (the day before the Ides). To make matters worse they counted inclusively so the 13th of March is III ante Idus and the 16th of March was XVII ante Kalendas Aprilis. And the duplicated day in leap years was VI ante Kalendas Martias.

The Gregorian calendar

THE JULIAN CALENDAR with its leap year every four years wasn't quite accurate enough since the year is not exactly 365.25 days long but about 365.2422. By the 16th century there was a discrepancy of about ten days and the vernal equinox was on March 11 rather than March 21.

In order to correct this Pope Gregory ordained that 5 October 1582 should be 15 October 1582. There was a bit of an outcry against the change because the ignorant people thought that they had some of their life stolen. It must have been particularly annoying for those with birthdays on the omitted days. And the other change was that from then on the century years such as 1600, 1700, should be leap years only when divisible by 400. So 1600 and 2000 were leap years as usual, but 1700, 1800, and 1900 were not.

The result is that there are 97 leap years in 400 years giving an average of 365.2425 days per year – an error of only 0.0003. A further correction will be needed in about 3000 years time.

The Gregorian calendar was immediately adopted by the Catholic countries in Europe such as France, Italy, Spain, and Portugal and the Catholic German states – they had no choice. But some countries took a long time to adjust. The exact years are not always clear but probably

> the Protestant German states, Netherlands, Denmark in 1700,
>
> Great Britain and colonies in 1752,
>
> Japan in 1872,
>
> China in 1912,
>
> Bulgaria in 1915,
>
> Turkey and Soviet Russia in 1918,
>
> Yugoslavia and Romania in 1919,
>
> Greece in 1923.

Of course, the later it was left the more adjustment was necessary. In the case of Britain, eleven days had to be omitted so that 3 September 1752 became 14 September 1752.

Sweden dithered. It omitted the leap day in 1700, and then had leap years in 1704 and 1708 but then had two leap days in 1708 thus reverting to the Julian calendar. It finally got sorted in 1753 when 18 February became 1 March.

Start of the year and quarters

THE FIRST MONTH of the Roman year was originally March but later became January. But there was much confusion until recent times. The idea of quarter days added to the problem.

Reckoning was often (and still sometimes is) done in terms of quarter days for purposes such as collecting rent. These are 25 March (Lady Day), 24 June (Midsummer Day), 29 September (Michaelmas), and 25 December (Christmas) – roughly the two solstices and equinoxes. In Scotland they used 2 February (Candelmas), 28 May (Whitsunday), 1 August (Lammas), and 11 November (Martinmas) although in 1990 they were changed to 28 February, May, August, and November but still keeping the traditional names – thus Lammas is now 28 August.

The Romans changed to use January as the first month so that the year began on 1 January. However, with the fall of the Roman Empire, confusion broke out. In the middle ages in England the year was reckoned to start at Christmas, but in the 12th century the Church decided that the year should begin on the feast of the Blessed Virgin or Lady Day which is March 25.

So the result was that there were two years, the legal year starting on 25 March and the historical year starting on 1 January. So a legal date in England of 1 March 1666 was the same as the historical date of 1 March 1667. Such dates were often written as 1 March 1666/7.

In England this confusion was eliminated with the adoption of the Gregorian calendar in 1752 and the year was then deemed to start on 1 January. But the tax collectors were unhappy with the prospect of a short year, so the tax year stayed the same and with the omission of eleven days, 25 March became 6 April. And so it has stayed ever since. Note that it was really only England (and colonies) that was a laggard regarding the start of the year. Other European countries including Scotland switched to 1 January much earlier.

This shift of eleven days was to some extent just absorbed into the quarter days. But contracts that finished on Lady Day had to be clarified as to whether they referred to the new 25 March or to 6 April which was sometimes referred to as Old Lady Day. Thus, in *Tess of the d'Urbervilles* by Thomas Hardy, poor Tess and her family are evicted on Old Lady Day which is 6 April.

It is annoying that the British tax year has not been aligned to start on 1 April (April Fool's Day) which would coincide with the VAT quarter.

The week

THE USE OF A WEEK of seven days goes back for at least 4000 years. It has no astronomical significance except perhaps as a handy subdivision of a month into about four weeks.

There are 52 weeks plus one or two days in a year. So a quarter is 13 weeks or so. Note that 90 days is often used as a legal period.

The names of the days of the week again reflect astronomical or mythological figures. In English they are of Germanic origin whereas in French they are typically Latin based.

Sunday   After the Sun naturally.
Monday   The Moon day. And similarly in French Lundi from Luna.
Tuesday   After Tyr, god of war. The French is Mardi from Mars, god of war.
Wednesday   Woden's day. The French is Mercredi from Mercury.
Thursday   Thor's day. The French is Jeudi from Jupiter.
Friday   From Frigga, goddess of love. The French is Vendredi, from Venus.
Saturday   From Saturn. The French is Samedi.

The French for Saturday (Samedi) and Sunday (Dimanche) have a somewhat different origin. The French words mean the day of the Sabbath and the day of the Lord. The Italian is clearer, Sabbato and Domenica. Also one wonders whether Frigga relates to frigging in the rigging!

Time of day

THE DAY is divided into 24 hours. It seems that this originated with the Egyptians who divided both day and night into 12 hours. The night hours were therefore longer in the winter. The reckoning of the hours was related to the rising of certain stars known as decans – there were 36 of these and they divided the year into 36 periods of 10 days each. The Greeks decided that having hours of different lengths was ridiculous and so they divided the day into 24 equal hours.

The hour is divided into 60 minutes and then into 60 seconds following the Babylonian system of using base 60.

Angles are similarly divided. A full circle is divided into 360 degrees, a degree is divided into 60 minutes and then into 60 seconds.

So a degree in angle is somehow correlated with an hour in time in that they both consist of 60 minutes. In terms of longitude, a difference of 15 degrees represents one hour difference in time.

Sunshine

THE SEASONS are caused by the fact that the axis of the Earth is inclined. As a consequence the period of daylight at typical latitudes is much less in the winter than in the summer. Moreover, the Sun rises less high in the sky in the winter and so sunlight strikes the Earth more obliquely at midday. As a consequence the energy from the Sun in the winter is very much less than in the summer. This effect gets worse as we go North since both the period of daylight gets less and the height of the Sun is less.

The effect in the summer is more complex. As we go further North, the period of daylight gets longer but the height of the Sun is less. The two effects cancel out to a remarkable degree.

We can relatively easily calculate the consequences at midsummer and midwinter. We make various assumptions such as the Earth is a perfect sphere of radius r and the orbit around the Sun is a circle. We also assume that the energy flux reaching the ground is not diminished by absorption when passing through the atmosphere (this is a bit optimistic perhaps but maybe reasonable on a clear day provided the Sun is not too low). This means that if the angle of the Sun to the vertical is φ, then the energy reaching the ground per unit area is proportional to $\cos \varphi$.

We assume that the inclination or tilt of the axis of the Earth is θ and the latitude of the point P concerned is λ. See the diagram opposite. (There is a hardish bit of maths coming up; if it seems daunting just skip a couple of pages to the equation marked D which is the key result and is then used to discuss the climate on the equator and at London and Stockholm.)

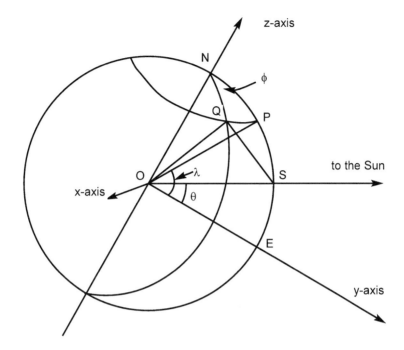

We take axes, x, y, z, as shown above. The axis of the Earth is the z-axis and so lies in the yz-plane which is the plane of the paper. N is the North pole. The x-axis comes straight out at us.

At midday the point P is also in the plane of the paper and the Sun S is at angle 90–(λ–θ) to the horizon or in other words at angle λ–θ to the vertical.

As the day progresses the Earth rotates. When it has rotated by an angle ϕ, the place which was originally at the point P at noon will have moved to the point Q. (Actually I have drawn the Earth rotating the wrong way but it doesn't matter.) For example at 2pm the angle ϕ will be 30°.

The point E is on the equator, O is the centre of the Earth and the Sun S can be assumed to be at infinity in the direction shown. The various angles are

$SOE = \theta$
$POE = \lambda$
$POS = \lambda-\theta$
$PNQ = \phi$
$QOS = \varphi$

QOS is the angle that the direction of the Sun is off the vertical when the place is at Q. Note that when Q was at P, then the angle φ was of course λ–θ.

As mentioned above, the energy reaching the Earth is diminished by the factor cos φ. If the Sun is directly overhead then φ is 0 and cos φ = 1 so that the energy is not diminished. When the Sun is on the horizon at sunset and sunrise then φ is 90° and since cos 90 = 0, no energy reaches the Earth at all.

108 Nice Numbers

So to find the total energy reaching the ground during a day we need to integrate cos φ from sunrise to sunset (or from sunrise to noon and double it). So we need to find a formula for φ in terms of the angle ϕ and then compute

$$H = \int_{\phi=0}^{\phi=F} \cos \varphi \, d\phi \qquad \ldots \text{(A)}$$

where F is the value of ϕ at sunrise (or sunset). Note that this gives the energy for just half a day.

Now consider the triangle QOS. It is isosceles with equal sides OS and OQ of length r. We need the angle $QOS = \varphi$. Half of this angle has a sine equal to half of QS divided by r. So

$$2 \sin \varphi/2 = QS/r$$

Now the points S and Q have spherical coordinates

$$S = (r, \theta, 0)$$
$$Q = (r, \lambda, \phi)$$

and so using the standard formulae

$$x = r \cos \theta \sin \phi$$
$$y = r \cos \theta \cos \phi$$
$$z = r \sin \theta$$

we find that the x, y, z coordinates are as follows

$$S = (0, r \cos \theta, r \sin \theta)$$
$$Q = (r \cos \lambda \sin \phi, r \cos \lambda \cos \phi, r \sin \lambda)$$

Now the square of the distance between two points (x, y, z) and (x', y', z') is

$$(x - x')^2 + (y - y')^2 + (z - z')^2$$

So applying this we have

$$QS^2 = r^2(\cos^2 \lambda \sin^2 \phi + (\cos \lambda \cos \phi - \cos \theta)^2 + (\sin \lambda - \sin \theta)^2)$$

But $QS^2 = 4r^2 \sin^2 \varphi/2$. So cancelling the common factor r^2, and multiplying all out, we have

$$4\sin^2 \varphi/2 = \cos^2 \lambda \sin^2 \phi + \cos^2 \lambda \cos^2 \phi - 2\cos \lambda \cos \phi \cos \theta + \cos^2 \theta$$
$$+ \sin^2 \lambda - 2 \sin \lambda \sin \theta + \sin^2 \theta$$
$$= \cos^2 \lambda - 2\cos \lambda \cos \phi \cos \theta + 1 + \sin^2 \lambda - 2 \sin \lambda \sin \theta$$
$$= 2 - 2\cos \lambda \cos \phi \cos \theta - 2 \sin \lambda \sin \theta$$

Now remember also the general formula

$$\cos 2x = 1 - 2\sin^2 x$$

So, putting $\varphi = 2x$, we have

$$2\sin^2 \varphi/2 = 1 - \cos \varphi$$

Now equating the two expressions for $2\sin^2\varphi/2$ (and cancelling a 2) and rearranging we finally get

$$\cos \varphi = \cos \lambda \cos \phi \cos \theta + \sin \lambda \sin \theta \qquad \ldots \text{(B)}$$

So at last we have a formula for $\cos \varphi$ in terms of ϕ (and λ and θ which are constants). The next thing we need is F which is the value of ϕ at sunrise (or sunset).

That is easy because $\cos \varphi$ is zero at sunrise and sunset so using the above equation (B), we immediately get

$$0 = \cos \lambda \cos F \cos \theta + \sin \lambda \sin \theta$$

So

$$\cos F = - \tan \lambda \tan \theta \qquad \ldots \text{(C)}$$

Now to do the integration to find the energy H. From equations (A and B) we have

$$H = \int_{\phi=0}^{\phi=F} \cos \varphi \, d\phi$$

$$= \int_{\phi=0}^{\phi=F} \cos \lambda \cos \phi \cos \theta + \sin \lambda \sin \theta \, d\phi$$

$$= \left[\cos \lambda \cos \theta \sin \phi + \sin \lambda \sin \theta \times \phi \right]_{\phi=0}^{\phi=F}$$

So finally, the total energy H received for half a day (from sunrise to noon or from noon to sunset) is given by the term for $\phi=F$ (that for $\phi=0$ is just zero)

$$H = \cos \lambda \cos \theta \sin F + \sin \lambda \sin \theta \times F \qquad \ldots \text{(D)}$$

We are now in a position to compute the energy received at midsummer and midwinter at various latitudes. The tilt of the Earth, θ, is 23.5°. So $\cos \theta = 0.917$ and $\sin \theta = 0.399$.

At the equator, $\lambda = 0$, so $\tan \lambda = 0$ also and using equation (C) we see that $\cos F$ is zero so that $F = 90°$ ($\pi/2$). This means that there are exactly 12 hours of daylight as expected on the equator. And then using equation (D) we have

$$H = \cos \theta = 0.917. \qquad \text{equator, midsummer}$$

We have done our calculation for midsummer. To get midwinter we change the sign of θ so that the Earth tilts the other way. But $\cos(-\theta) = \cos \theta$ so it makes no difference on the equator. So midsummer and midwinter are the same on the equator as we would expect.

We can also use our results for the equinoxes since so far as energy is concerned it is exactly as if there were no tilt. So putting $\theta = 0$ we have

$$H = \cos 0 = 1.000 \qquad \text{equator, equinox}$$

That's interesting – the equator gets more sunshine at the equinoxes than at midsummer.

Let's try London, where $\lambda = 51.5$. Now $\tan 51.5 = 1.258$ and $\tan 23.5 = 0.43$ so we find from equation (C) that

$$\cos F = -\tan \lambda \tan \theta = -0.546$$

$$F = 123°$$

That seems correct. That means London gets 246 out of 360 degrees of sunshine which is 16 hours and 24 minutes which is very close to the correct amount. Putting these values in equation (D) for H we have

$$\begin{aligned} H &= \cos \lambda \cos \theta \sin F + \sin \lambda \sin \theta \times F \\ &= 0.623 \times 0.917 \times 0.839 + 0.782 \times 0.399 \times 2.146 \\ &= 0.479 + 0.667 = \\ &= 1.147 \qquad \text{London, midsummer} \end{aligned}$$

So London actually gets more energy from the Sun at midsummer than on the equator. The fact that the Sun is not so high in the sky is more than compensated for by the longer hours of daylight.

If we do the same calculation for London in midwinter then we find that F is only $57°$ and θ is now -23.5. We get

$$\begin{aligned} H &= 0.623 \times 0.917 \times 0.839 - 0.782 \times 0.399 \times 0.995 \\ &= 0.479 - 0.310 \\ &= 0.169 \qquad \text{London, midwinter} \end{aligned}$$

Brrr! In midwinter in London we get only about $1/7$th of the energy from the Sun that we get in midsummer. No wonder it seems so miserable!

At the equinoxes, $F = 90°$ and we take $\theta = 0$. Most terms become 1 or 0 and we get

$$H = \cos \lambda = 0.623 \qquad \text{London, equinox}$$

So that's not too bad. Slightly more than half of midsummer and over three times as much as midwinter.

Finally, we do the calculations for the North Pole where $\lambda = 90°$. At midsummer the Sun does not set and so we take $F = 180° (= \pi)$. The expression for H (from equation D) simplifies since $\sin F = 0$, $\cos F = 1$ and $\sin \lambda = 1$. We get

$$H = \pi \sin \theta = 3.142 \times 0.399 = 1.253$$

That's amazing. At midsummer the North Pole actually gets more energy from the Sun than anywhere else on the planet (ignoring absorption by clouds).

Incidentally, note that we should really have multiplied all the figures by 2 since we only did the calculation from sunrise to midday. However, it seems nice to have a standard figure of 1.0 for the equator at the equinoxes.

The first table overleaf shows the corresponding figures for various latitudes. The figures for midsummer are rather surprising. As we go North the amount of energy steadily increases until at latitude 44° it has a local maximum, and then reduces slightly with a local minimum at latitude 62° and it then increases to a maximum at the North Pole.

The local maximum at latitude 44° corresponds to places such as Nice and Florence. Maybe that is why the classic civilizations emerged around the Mediterranean.

Stockholm with latitude about 60° is close to the local minimum and has figures as follows

$H = 1.139$ Stockholm, midsummer

$H = 0.054$ Stockholm, midwinter

$H = 0.500$ Stockholm, equinox

So Stockholm has less than half the sunshine than London in midwinter and a shade less than London in midsummer.

The other tables show the results for different angles of tilt. No tilt at all is boring. And the equinox is generally boring being independent of the tilt.

The general picture in midsummer of getting more sunshine as we go North and then a dip before going up again is present at 10° and 20° but not at 30°. It stops at around 25°. Above this it just keeps increasing to a maximum at the North pole. With a tilt of 90° the North pole gets π times more than the equator at the equinox.

It looks as if the tilt of the Earth is fairly optimal. We have decent seasons without being too extreme at high latitudes.

The reader might like to consider how the calculations might be extended to cover any time of year. And then to calculate the total energy over the whole year at various latitudes.

112 Nice Numbers

Latitude	Winter	Equinox	Summer
0.0	0.917	1.000	0.917
2.0	0.895	0.999	0.938
4.0	0.872	0.998	0.959
6.0	0.848	0.995	0.978
8.0	0.823	0.990	0.997
10.0	0.797	0.985	1.015
12.0	0.771	0.978	1.031
14.0	0.744	0.970	1.047
16.0	0.716	0.961	1.061
18.0	0.687	0.951	1.074
20.0	0.658	0.940	1.087
22.0	0.629	0.927	1.098
24.0	0.599	0.914	1.108
26.0	0.568	0.899	1.117
28.0	0.537	0.883	1.126
30.0	0.506	0.866	1.133
32.0	0.475	0.848	1.139
34.0	0.443	0.829	1.143
36.0	0.411	0.809	1.147
38.0	0.379	0.788	1.150
40.0	0.347	0.766	1.152
42.0	0.315	0.743	1.154
44.0	0.284	0.719	1.154
46.0	0.252	0.695	1.153
48.0	0.221	0.669	1.152
50.0	0.191	0.643	1.150
52.0	0.161	0.616	1.148
54.0	0.132	0.588	1.146
56.0	0.104	0.559	1.143
58.0	0.078	0.530	1.140
60.0	0.054	0.500	1.139
62.0	0.032	0.469	1.138
64.0	0.014	0.438	1.140
66.0	0.001	0.407	1.146
68.0	0.000	0.375	1.161
70.0	0.000	0.342	1.177
72.0	0.000	0.309	1.191
74.0	0.000	0.276	1.204
76.0	0.000	0.242	1.215
78.0	0.000	0.208	1.225
80.0	0.000	0.174	1.234
82.0	0.000	0.139	1.241
84.0	0.000	0.105	1.246
86.0	0.000	0.070	1.250
88.0	0.000	0.035	1.252
90.0	0.000	0.000	1.253

Table above shows sunshine at various latitudes on Earth at midsummer, midwinter, and the equinoxes. Table opposite shows the effect of tilt.

5 Time   113

Tilt = 0.0				Tilt = 50.0			
Latitude	Winter	Equinox	Summer	Latitude	Winter	Equinox	Summer
0.0	1.000	1.000	1.000	0.0	0.643	1.000	0.643
10.0	0.985	0.985	0.985	10.0	0.438	0.985	0.856
20.0	0.940	0.940	0.940	20.0	0.250	0.940	1.073
30.0	0.866	0.866	0.866	30.0	0.093	0.866	1.296
40.0	0.766	0.766	0.766	40.0	0.000	0.766	1.547
50.0	0.643	0.643	0.643	50.0	0.000	0.643	1.844
60.0	0.500	0.500	0.500	60.0	0.000	0.500	2.084
70.0	0.342	0.342	0.342	70.0	0.000	0.342	2.261
80.0	0.174	0.174	0.174	80.0	0.000	0.174	2.370
90.0	0.000	0.000	0.000	90.0	0.000	0.000	2.407

Tilt = 10.0				Tilt = 60.0			
Latitude	Winter	Equinox	Summer	Latitude	Winter	Equinox	Summer
0.0	0.985	1.000	0.985	0.0	0.500	1.000	0.500
10.0	0.923	0.985	1.018	10.0	0.279	0.985	0.752
20.0	0.834	0.940	1.021	20.0	0.101	0.940	1.032
30.0	0.721	0.866	0.994	30.0	0.000	0.866	1.360
40.0	0.587	0.766	0.938	40.0	0.000	0.766	1.749
50.0	0.438	0.643	0.856	50.0	0.000	0.643	2.084
60.0	0.279	0.500	0.752	60.0	0.000	0.500	2.356
70.0	0.121	0.342	0.633	70.0	0.000	0.342	2.557
80.0	0.000	0.174	0.537	80.0	0.000	0.174	2.679
90.0	0.000	0.000	0.546	90.0	0.000	0.000	2.721

Tilt = 20.0				Tilt = 70.0			
Latitude	Winter	Equinox	Summer	Latitude	Winter	Equinox	Summer
0.0	0.940	1.000	0.940	0.0	0.342	1.000	0.342
10.0	0.834	0.985	1.021	10.0	0.121	0.985	0.633
20.0	0.707	0.940	1.075	20.0	0.000	0.940	1.010
30.0	0.563	0.866	1.100	30.0	0.000	0.866	1.476
40.0	0.408	0.766	1.099	40.0	0.000	0.766	1.898
50.0	0.250	0.643	1.073	50.0	0.000	0.643	2.261
60.0	0.101	0.500	1.032	60.0	0.000	0.500	2.557
70.0	0.000	0.342	1.010	70.0	0.000	0.342	2.774
80.0	0.000	0.174	1.058	80.0	0.000	0.174	2.907
90.0	0.000	0.000	1.074	90.0	0.000	0.000	2.952

Tilt = 30.0				Tilt = 80.0			
Latitude	Winter	Equinox	Summer	Latitude	Winter	Equinox	Summer
0.0	0.866	1.000	0.866	0.0	0.174	1.000	0.174
10.0	0.721	0.985	0.994	10.0	0.000	0.985	0.537
20.0	0.563	0.940	1.100	20.0	0.000	0.940	1.058
30.0	0.399	0.866	1.185	30.0	0.000	0.866	1.547
40.0	0.238	0.766	1.248	40.0	0.000	0.766	1.989
50.0	0.093	0.643	1.296	50.0	0.000	0.643	2.370
60.0	0.000	0.500	1.360	60.0	0.000	0.500	2.679
70.0	0.000	0.342	1.476	70.0	0.000	0.342	2.907
80.0	0.000	0.174	1.547	80.0	0.000	0.174	3.047
90.0	0.000	0.000	1.571	90.0	0.000	0.000	3.094

Tilt = 40.0				Tilt = 90.0			
Latitude	Winter	Equinox	Summer	Latitude	Winter	Equinox	Summer
0.0	0.766	1.000	0.766	0.0	0.000	1.000	0.000
10.0	0.587	0.985	0.938	10.0	0.000	0.985	0.546
20.0	0.408	0.940	1.099	20.0	0.000	0.940	1.074
30.0	0.238	0.866	1.248	30.0	0.000	0.866	1.571
40.0	0.090	0.766	1.388	40.0	0.000	0.766	2.019
50.0	0.000	0.643	1.547	50.0	0.000	0.643	2.407
60.0	0.000	0.500	1.749	60.0	0.000	0.500	2.721
70.0	0.000	0.342	1.898	70.0	0.000	0.342	2.952
80.0	0.000	0.174	1.989	80.0	0.000	0.174	3.094
90.0	0.000	0.000	2.019	90.0	0.000	0.000	3.142

Further reading

A GOOD DISCUSSION of the calendar will be found in *Whittaker's Almanac*. Not only does that cover the Roman (Julian) and later Gregorian calendars and the leap year problem but also discusses the Metonic cycle of 19 years which forms the basis of many other calendars.

See also https://en.wikipedia.org/Julian_Calendar which casts doubt on the story about the emperor Augustus.

A more comprehensive discussion will be found in *The Calendar* by David Duncan.

Exercises

1. A man in London sends a letter to his mistress in Paris and she replies immediately. The man dates his letter 1st February 1720. Assuming that it takes five days to reach his mistress, what is the date on her reply?
2. What would the tilt of the Earth have to be if the sunshine at the North Pole at Midsummer were 1.0 (the same as at the Equator at an Equinox)?
3* Find a general formula for the amount of sunshine received during a day at any time of the year at any latitude. And then integrate to find the total over a whole year. You are permitted to do the integration by a numerical method.

6 Notations

THIS LECTURE LOOKS at the different ways in which numbers can be expressed and the importance of zero. As examples it considers the Egyptian, Babylonian, and Roman systems. It also considers representations in different bases and tests for divisibility. It concludes by encountering Fermat's Little Theorem involving prime numbers and modular arithmetic.

Types of notation

THERE ARE two basic ways in which values are expressed. In one way the value of a symbol depends upon its place in the number. We call these place notations. Our own decimal notation is of this form. When we write 474, this means

$$(4 \times 100) + (7 \times 10) + (4 \times 1) = 400 + 70 + 4$$

The first 4 has value 400 but the second 4 has value of just 4. The value of the 4 depends upon it place. That's about all there is to it.

If a place has nothing in it then we put zero to indicate this. We use base 10 so that the value of a digit is a power of 10 thus

$$4609 = 4 \times 10^3 + 6 \times 10^2 + 0 \times 10^1 + 9 \times 10^0 = 4000 + 600 + 9$$

Remember that we have the natural convention that anything raised to the power 0 is simply 1.

We can use other bases such as 2 (binary) and 8 (octal). We will come back to the use of other bases later.

The symbols used here, namely 0, 1, 2, 3, 4, 5, 6, 7, 8, 9 which we can refer to as Western Arabic were derived from an older Arabic form which is used in conjunction with many languages today. These symbols are shown below. It might be thought that there could be confusion between the symbols for 7 and 8 since one is the inverse of the other. But then 6 is equally the inverse of 9 and indeed they are sometimes confused.

0 1 2 3 4 5 6 7 8 9

٠ ١ ٢ ٣ ٤ ٥ ٦ ٧ ٨ ٩

Arabic numerals.

Nice Numbers

The iconic camel mail stamps. Note the gait of the camel. This will be discussed in the final lecture.

Many countries have postage stamps with values in both forms. The iconic camel mail stamps from the Sudan are typical and are illustrated above.

The other basic type of notation is where the value of a symbol is the same wherever it occurs in a number. A good example is the Egyptian system which uses different symbols for 1, 10, 100, and so on as shown in the diagram below. Thus 1 is represented by a simple stroke, 10 by a sort of hoop, 100 by a coil of rope, 1000 by a lotus flower, 10,000 by a bent finger, and 100,000 by a tadpole.

General numbers are then represented by a suitable combination thus

|||∩   13

|૭   101

Even though the value of a symbol was not determined by its place, nevertheless the Egyptians always wrote the symbols in descending order of value and so, remembering that they wrote from right to left, the lowest value symbol would be on the left. Moreover, if there were more than four of one symbol then they would be written as two rows.

So a number such as 365 would be

||| ∩∩∩ ૭૭૭   365
|| ∩∩∩

There is of course no need for a symbol for zero. Thus 2012 is written as

||∩ ɭɭ   2012

and we do not have to indicate that there are no coils since it is perfectly obvious that there are none.

1	│	stroke
10	∩	hoop
100	૭	coil
1000	ɭ	lotus
10000	╲	finger
100000	⌒	tadpole

Egyptian hieroglyphic symbols for powers of 10.

Roman numbers are somewhat similar. Thus when we write

    MMVII   2007

the two Ms are both of value 1000 and the two Is are both of value 1. The fact that we have no hundreds or tens does not shown up and there is no need for a symbol for zero.

Curiously enough Roman numbers have survived in our culture for certain uses so we are reasonably familiar with them. They have interesting rules which we will now look at in some detail.

Roman numbers

ROMAN NUMBERS are basically written as a sequence of Roman digits whose values are as follows

    I   1
    V   5
    X   10
    L   50
    C   100
    D   500
    M   1000

Although the value of a Roman digit does not depend upon where it is placed, there is a convention in writing Roman numbers (similar to that for Egyptian numbers) that we always put the higher valued digits first. However, there is an additional rule that if a single lower symbol precedes one of higher value then the lower is subtracted from the higher, thus

    IV   4
    IX   9
    XL   40
    XC   90
    CD   400
    CM   900

Note that this rule only applies to the situations shown. One should not write IC for 99 but should write it out in detail thus

    XCIX   99

But sometimes one encounters a violation of the rule. There is a statue in Pisa with the value VC for an age. Clearly it is not the Victoria Cross but 95.

118 Nice Numbers

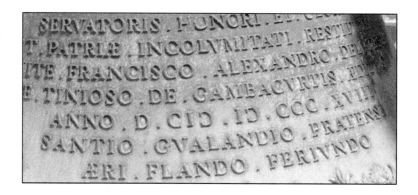

A bell in the Leaning Tower of Pisa.

The origin of the value of the letters is mixed and some are a bit obscure.

I for 1 is undoubtedly just a tally mark. V is perhaps half of X where X is another tally mark. This is similar to the way we often make single marks when counting and then a line through to complete a group of five as shown below.

The origins of C and M are straightforward. C is short for Centum, Latin for a hundred and M is for Mille, Latin for a thousand.

L and D are more obscure. The origin of L seems to be unknown. And D is curious. There was an earlier notation for 1000 and 500 thus

    CIƆ   1000
    IƆ   500

The photograph above shows the use of these on a bell in the Leaning Tower of Pisa. The line starting Anno reads

    ANNO . D . CIƆ . IƆ . CCC . XVIII

The first D is simply for Domini as part of Anno Domini. The date itself appears to be 1000+500+300+18 = 1818. However, this does not seem consistent with published records regarding the bells.

Why do we still use Roman numbers for certain applications? Probably because they lend an air of majesty.

An important application is in clock faces. Many clocks and watches use Roman numbers. But note an important variation. It is almost always the case

Tally marks for 13.

Clocks with Roman numerals usually use IIII rather than IV.

that 4 is not represented by IV but by IIII. It seems that this is in order to help to balance the appearance of the left and right sides of the dial as illustrated above.

Another application is in dates. The dates at the ends of films and TV shows are almost always given as Roman numbers. But copyright and printing dates are now rarely given as Roman numbers in books. An amusing example of a date is obtained by putting them in descending order thus

    MDCLXVI   1666

which is the date of the Great Fire of London! Moreover, DCLXVI is the number of the Beast!

Roman numbers are often used for numbering auxiliary pages. Sometimes a newspaper supplement uses Roman numbers and preliminary pages in a book almost always use Roman numbers. Other applications in books include chapter numbers, appendix numbers, diagram and figure numbers and so on.

The disadvantage of Roman numbers is that they are hard to manipulate. Thus try multiplying XXI by IV without first converting them to base 10 place notation. We note that the applications mentioned above do not involve manipulation but simply enumeration. It does not make any sense to multiply a year by a page number.

There were some variations on the normal rules. Very large numbers could be formed by placing a bar above a symbol to mean 1000 times, so

    $\bar{M}$ one million
    $\bar{V}$ five thousand

Also the backward C could be used to mean multiply by ten, thus

    LƆƆ   five thousand

These variations are rarely encountered.

Roman numbers are often written in lower case. Sometimes the letter j is used instead of i as the last letter thus

    xxiij   23

It is believed that this was done to prevent forgery by the addition of an extra i. But clearly it didn't help if the last letter was not an i!

Babylonian system

THE BABYLONIANS evolved what was essentially a place system using base 60. That sounds as if it would need 60 different symbols for individual digits. But they had a two tier system which had distinct symbols for 1 and 10. These were made by pressing a reed into wet clay and the symbols were something like

    𒁹   1

    𒌋   10

Numbers up to 59 were made by combinations of these, thus

    𒌋𒌋𒁹𒁹𒁹   23

Numbers such as 9 were made by compacting together several of the symbols for 1 but that is a minor detail.

The important point is that they represented larger numbers using base 60. They used extra space to separate the individual base 60 digits. Here are some examples

    𒁹 𒌋𒌋𒌋𒌋𒁹 $101 = 1 \times 60 + 41$

    𒁹𒁹𒁹 𒁹𒁹𒁹 $183 = 3 \times 60 + 3$

    𒁹 𒁹𒁹 $62 = 1 \times 60 + 2$

    𒁹 𒁹 𒁹 $3661 = 1 \times 60^2 + 1 \times 60 + 1$

Note the possibility for enormous confusion regarding the placing of the spaces. Numbers written in a hurry were often confused.

And there was much greater confusion when they wanted to write 3601 which of course is

    𒁹 𒁹 $3601 = 1 \times 60^2 + 0 \times 60 + 1$

They had no symbol for zero and maybe they tried to indicate that a place was unused by leaving a larger gap. Eventually they used a sort of sloping 2 as a kind of separator. This stood for zero in the middle of a number thus

    𒁹 ⌃ 𒁹   3601

However, they did not use this at the end of a number to represent zero, so if they wrote

    ⊏⊏ 𒁹   21 or 1260?

then it was ambiguous and they had to rely upon context. I suppose that with a base as large as 60 one would hardly accidentally purchase 1260 cows rather than just 21 cows.

We have retained the base 60 system for times and angles and we distinguish the places by appending an appropriate symbol thus

    37°6'27"   37 degrees, 6 minutes, and 27 seconds

or, in the case of times, a colon is often used as a separator thus 14:08:30.

Place systems and bases

A PLACE SYSTEM only really becomes unambiguous if a) every different value in the base range has a distinct symbol, and b) there is a symbol for zero. This only really happened around the 9th century in India and slowly made its way to Europe.

Having a distinct symbol for every value in the base range is difficult if the base is large. Thus having 60 different symbols to represent 0 to 59 would be tricky. And as we have just seen the Babylonians used a double system which overcomes this dificulty.

In computing, numbers are often written in hexadecimal (base 16) and the values 10 to 15 are represented by the letters A to F.

    ACE $10 \times 16^2 + 12 \times 16 + 14 = 2766$
    BED $11 \times 16^2 + 14 \times 16 + 13 = 3053$
    FACE $15 \times 16^3 + 10 \times 16^2 + 12 \times 16 + 13 = 64{,}205$

Advocates of a duodecimal number system (base 12) have suggested t and e (or variations of them) for ten and eleven thus

    2et2 $2 \times 12^3 + 11 \times 12^2 + 10 \times 12 + 2 = 5162$

Other commonly used bases in computing are binary (base 2) and octal (base 8) as follows

    10011   19   binary
    23   19   octal
    111111   63   binary
    77   63   octal

122 Nice Numbers

Binary numbers are tedious to read because it is hard to keep track of the many digits that naturally arise. As a consequence octal and hexadecimal numbers are frequently used. When it is necessary to distinguish the base we can write the base value as a sort of subscript thus

111_2	7	$= 4 + 2 + 1$
111_5	31	$= 25 + 5 + 1$
111_8	73	$= 64 + 8 + 1$
111_{10}	111	$= 100 + 10 + 1$
111_{16}	273	$= 256 + 16 + 1$

The rules for arithmetic in any base are really the same. It is just that the values carried are different.

Conversion from one base to another is done by division. Thus to convert 123 in base 10 into octal we keep dividing by 8 and noting the remainder until the quotient is less than 8 thus

$123 \div 8 = 15$ remainder 3
$15 \div 8 = 1$ remainder 7

The final quotient and the remainders then form the answer (in red) thus

173_8 $(1 \times 64) + (7 \times 8) + (3 \times 1) = 64 + 56 + 3 = 123$

And to convert 173 in octal to decimal we keep dividing by 10 (being careful to remember that we are working in octal so we should write the 10 as 12)

We get

$173 \div 12 = 14$ remainder 3
$14 \div 12 = 1$ remainder 2

and so we have got back to 123 in decimal.

This is quite tricky to do. To work out $173 \div 12$ in base 8 we really need to do long division. Remember how we would do $247 \div 14$ in normal decimal notation.

As shown in the first diagram opposite, 14 goes once into 24, so the first digit of the answer is 1 and the remainder is 10. We then pull down the 7 to give the next dividend of 107. Now 14×7 is 98 so the next digit is 7 and remainder 9. There are no more digits to pull down so we are done.

We do exactly the same in base 8. As shown in the second diagram, 12 goes once into 17, so the first digit of the answer is 1 and the remainder is 5. We then pull down the 3 to give the next dividend of 53. Now 12×4 is 50 in base 8 so the next digit is 4 and the remainder is 3. There are no more digits to pull down so we are done.

```
    17
14)247          Dividing 247 by 14        14       Dividing 173 by 12
 14             in base 10 notation    12)173      in base 8 notation
---             by long division.         12       by long division.
107                                      ---
 98             The answer is 17          53       The answer is 14
---             with remainder 9.         50       with remainder 3.
  9                                      ---
                                           3
```

We really need to write out the base 8 times table as shown below. Many of the entries seem familiar, thus $12 \times 12 = 144$ in all bases above 4 since no carrying is involved. The table is extended to 12 times (that is 10 in decimal) just as we extend the 10 times table to 12 for convenience. This probably reflects the use of duodecimals as in feet and inches.

Another example might help. Consider 4321 in base 10 and convert to octal thus

$4321 \div 8 = 540$ rem **1**
$540 \div 8 = 67$ rem **4**
$67 \div 8 = 8$ rem **3**

So the value in octal is 8341. And then we convert back to check by dividing by 10 (that is 12 in octal) with the help of the 8 times table thus

$8341 \div 12 = 660$ rem **1**
$660 \div 12 = 53$ rem **2**
$53 \div 12 = 4$ rem **3**

And we have 4321 back again.

Bases other than 2, 8, 12, and 16 (and of course 10) are rarely encountered but are sometimes used for amusement.

1	2	3	4	5	6	7	10	11	12
2	4	6	10	12	14	16	20	22	24
3	6	11	14	17	22	25	30	33	36
4	10	14	20	24	30	34	40	44	50
5	12	17	24	31	36	43	50	55	62
6	14	22	30	36	44	52	60	66	74
7	16	25	34	43	52	61	70	77	106
10	20	30	40	50	60	70	100	110	120
11	22	33	44	55	66	77	110	121	132
12	24	36	50	62	74	106	120	132	144

The base 8 table.

We go up to 12 times 12 for convenience as we do for the decimal table.

A present from the White Elephant showing the use of base 4.

The photograph above shows a celebratory corkscrew commemorating 100_4 or 16_{10} years of the White Elephant software company in Switzerland.

Divisibility

AT THIS POINT the reader might care to look back at the section on modular arithmetic in the lecture on Amicable Numbers. Remember that the basic idea is that we do arithmetic and cast out multiples of some modulus after each operation. We say that two numbers a and b are congruent with respect to a base m if $a-b$ is exactly divisible by m. We use the symbol ≡ rather than = in order to emphasize the difference from normal equality. In summary, the key definition is

$a \equiv b \pmod{m}$ means $a-b$ is exactly divisible by m

Remember that the usual rules always work for addition, subtraction, and multiplication but not for division.

Congruencies are helpful in understanding divisibility rules. For example, in order to find out whether a number is divisible by 3 we just add up its digits and if that sum is divisible by 3 then the original number is divisible by 3 as well.

Thus consider 12345. Its digits add up to 15 (1+2+3+4+5 = 15) and 15 is exactly divisible by 3 and so 12345 is also divisible by 3. In fact 12345 = 4115 × 3. Why does this work?

Now

$1 \equiv 1 \pmod{3}$
$10 \equiv 1 \pmod{3}$

and since we know that multiplication preserves congruence we can keep multiplying by 10 on the left and 1 on the right thus

$100 \equiv 1 \pmod{3}$
$1000 \equiv 1 \pmod{3}$

Now a number such as 12345 means

$$1 \times 10000 + 2 \times 1000 + 3 \times 100 + 4 \times 10 + 5 \times 1$$

and now we can apply congruence mod 3 throughout and replace all the powers of 10 by 1 since they are all congruent to 1. Obviously we get

$$1 + 2 + 3 + 4 + 5$$

and so if this is equal to zero mod 3 (which means it is divisible by 3 exactly) then the original number is also divisible by 3. The digits add to 15 which is divisible by 3 and so 12345 is also divisible by 3.

In the case of starting with a large number we can continue the process of adding the digits until we get a single digit. Thus we add the digits of 15 and get 6. And of course 6 is also divisible by 3. This final single digit is often known as the digital root of the original number. In his famous book *Amusements in Mathematics*, Dudeney describes a number of puzzles using digital roots.

An example is illustrated by Puzzle 76. A man buys six barrels. One has beer and the other five have wine. The volumes are 15, 16, 18, 19, 20, and 31 gallons. He sells some barrels to one friend and some others (twice the volume) to another friend and keeps the one barrel with the beer for himself. Which one has the beer?

Now the digital roots of the six numbers are 6, 7, 9, 1, 2, and 4. And the digital root of the sum of these is 2 so the total volume has remainder 2 when divided by 3. However, since the total volume sold must divide exactly by 3 (since one sale was precisely twice the other), it follows that the barrel of beer must have a volume with remainder 2 when divided by 3. The only barrel with that property is that of 20 gallons so that one contains the beer.

Let's return to divisibility and explore the divisibility test for 11. By repeated multiplication by 10 we get

$$1 \equiv 1 \pmod{11}$$
$$10 \equiv -1 \pmod{11}$$
$$100 \equiv 1 \pmod{11}$$
$$1000 \equiv -1 \pmod{11}$$
$$10000 \equiv 1 \pmod{11}$$
$$\ldots$$

So alternate digits are multiplied by $+1$ and -1. So to test 12345 for divisibility by 11 we get

$$1 - 2 + 3 - 4 + 5 = 3$$

and so the remainder on dividing 12345 by 11 is 3 and so it is not divisible by 11. Observe that $12345 = 1122 \times 11 + 3$.

126   Nice Numbers

We can do the same with divisibility by 7 although it is not so easy. We have

$1 \equiv 1 \pmod{7}$
$10 \equiv 3 \pmod{7}$
$100 \equiv 2 \pmod{7}$ since $3 \times 3 = 9 \equiv 2 \pmod{7}$
$1000 \equiv 6 \pmod{7}$ since $2 \times 3 = 6$
$10000 \equiv 4 \pmod{7}$ since $6 \times 3 = 18 \equiv 4 \pmod{7}$
$100000 \equiv 5 \pmod{7}$ since $4 \times 3 = 12 \equiv 5 \pmod{7}$
$1000000 \equiv 1 \pmod{7}$ since $5 \times 3 = 15 \equiv 1 \pmod{7}$
...

and then the sequence repeats. Now we can apply the test to 12345 and get

$$(1 \times 4) + (2 \times 6) + (3 \times 2) + (4 \times 3) + (5 \times 1) =$$
$$4 + 12 + 6 + 12 + 5 = 39 \equiv 4 \pmod{7}$$

and so the remainder on dividing 12345 by 7 is 4. Indeed $12345 = 1763 \times 7 + 4$ so it works.

But that test for divisibility by 7 is fairly tedious – we might as well just divide the wretched number by 7 in the first place to find the remainder.

This whole mechanism works in other bases as well. Consider base 8 and let's devise a test for divisibility by 3.

$1 \equiv 1 \pmod{3}$
$10_8 \equiv -1 \pmod{3}$
$100_8 \equiv 1 \pmod{3}$
$1000_8 \equiv -1 \pmod{3}$
$10000_8 \equiv 1 \pmod{3}$
...

So the test for divisibility by 3 in base 8 is like the test for divisibility by 11 in base 10. Suppose we convert 12345 in base 10 into base 8. We get 30071_8. and then

$$3 - 0 + 0 - 7 + 1 = -3$$

and –3 is divisible by 3 so it works.

We can devise a test for divisibility by 7 in base 8 in the same way. We get

$1 \equiv 1 \pmod{7}$
$10_8 \equiv 1 \pmod{7}$
$100_8 \equiv 1 \pmod{7}$
...

and this shows that the test for divisibility by 7 in base 8 is very easy – we simply add up the digits. So for 30071_8 we get the sum of the digits is 11 which is congruent to 4 mod 7 and so the remainder is 4 which confirms the previous result. However, although this divisibility test is very easy, the conversion of the number from decimal digits to base 8 is fairly tedious – it really is easier just to divide the original number by 7.

The divisibility test for numbers such as 101 and 1001 is much like that for 11 but uses groups of digits. Thus consider 101. We can consider a number expressed in decimal digits to be a number in base 100 by simply taking the digits in pairs. Thus 123456 can be considered as

$$12 \times 100^2 + 34 \times 100 + 56 \times 1$$

So we can devise a divisibility test by 101 by noting that

$1 \equiv 1 \pmod{101}$
$100 \equiv -1 \pmod{101}$
$10000 \equiv 1 \pmod{101}$
$1000000 \equiv -1 \pmod{101}$
...

To test for divisibility by 11 we add and subtract alternate digits. For 101 we add and subtract alternate pairs of digits. Let's try it on 12,469,056 which we can subdivide as

12'46'90'56.

Now consider $(56+46) - (90+12) = 0$. That means that 12,469,056 is divisible by 101. And indeed it equals $101 \times 123,456$.

We can do a similar test for divisibility by 1001. We have

$1 \equiv 1 \pmod{1001}$
$1,000 \equiv -1 \pmod{1001}$
$1,000,000 \equiv 1 \pmod{1001}$
$1,000,000,000 \equiv -1 \pmod{1001}$
...

We just group the digits in threes and then add and subtract these groups. Consider 123,579,456. We mark it as

123'579'456.

which is very natural because we often mark large numbers into threes by using commas anyway. And then consider 456+123−579. This is zero and so 123,579,456 is divisible by 1001.

Divisibility by 1001 is not in itself very interesting but note carefully that $1001 = 7 \times 11 \times 13$. We can therefore use this test to see whether a number is divisible by 7 or 13 (or indeed 11 but that's easy anyway).

n	$10^{n-1}+1$	factors
2	11	11
3	101	101
4	1001	7.11.13
5	10001	73.137
6	100001	11.9091
7	1000001	101.9901
8	10000001	11.909091
9	100000001	17.5882353
10	1000000001	7.11.13.19.52579
11	10000000001	101.3541.27961
12	100000000001	11^2.23.4093.8779

Factors of 11, 101, 1001 etc.

Consider 864,197,523. We mark it as 864'197'523. Now 523+864–197 = 1190. Since 1190 is divisible by 7 (it equals 170×7), it follows that 864,197,523 is also divisible by 7. That is a better test for divisibility by 7.

Similarly, consider 1,604,938,257. We have 1+938–604–257 = 78. Since 78 is divisible by 13, it follows that 1,604,938,257 is also divisible by 13. It is in fact 13×123456789.

What about 10001? It is 73×137 so we can devise a similar test for divisibility by 73 or 137 by grouping the digits into blocks of four. Consider 9,012,345,597. Mark it into fours thus 90'1234'5597 and consider 90+5597–1234 = 4453. Now 4453 is divisible by 73 and so 9,012,345,597 is also divisible by 73.

We therefore see that numbers of the form 10...01 have interesting factors as do factors of 11...11 which we met when considering recurring decimals in the lecture on Fractions. There is a strong relationship between the factors of these numbers because of facts such as 111111 = 111 × 1001.

The table above shows the factors and we see that we could devise a test for divisibility by 17 by grouping digits into blocks of 8, and for 19 by grouping digits into blocks of 9. Hardly worth the effort though!

Fractions and bases

WE WILL NOW consider fractions in various bases other than 10. The values of the digits after the point represent an inverse fraction of powers of the base. If the base is n then the value of 0.1 is of course $1/n$ and 0.01 is $1/n^2$ and so on.

In base 12 for example we have

$1/2 = 0.6$ in base 12
$1/3 = 0.4$
$1/4 = 0.3$
$1/6 = 0.2$
$1/8 = 0.16$

and we see that 1/3 which is a recurring fraction in base 10 does not recur in base 12. That is because 3 is a factor of 12 and indeed 1/3 equals 4/12.

On the other hand 1/5 does recur in base 12

$1/5 = 0.24972497...$

and the cycle length is 4. Recall from the lecture on Fractions that the cycle length for $1/n$ (where n is prime) must be a factor of $n-1$. So 1/5 in base 12 has its maximum cycle length.

We might wonder what the value of 0.1111... is in base 12. We get

$x = 0.1111...$
$12x = 1.1111...$ and then subtracting
$11x = 1$
$x = 1/11$

We should really have done that using base 12 notation throughout thus

$x = 0.1111....$
$10x = 1.1111...$
$ex = 1$
$x = 1/e$

where we use e to mean the digit for eleven. So the fraction for $1/11_{12}$ is simply a recurring sequence of 1s. In fact in any base b, the fraction $1/(b-1)$ is a sequence of 1s.

The sequence for 1/7 was nasty in decimals. In base 12 it is

$1/7 = 0.186t35186t35...$

and again of length 6. Note that we use t to mean the digit for ten.

1/13 was also rather nasty in decimal but in base 12 we get

$1/13 = 0.0e0e0e...$

which echoes 1/11 in decimal. The sequence is simply zero and one less than the base repeatedly.

	base 12	decimal	factors	in base 12
1	1			
2	11	13	13	11
3	111	157	157	111
4	1111	1885	5.13.29	5.11.25
5	11111	22621	22621	11111
6	111111	271453	7.13.19.157	7.11.17.111
7	1111111	3257437	659.4943	46e.2t3e
8	11111111	39089245	5.13.29.89.233	5.11.25.75.175

Factors of 11, 111, 1111 etc in base 12.

The reader might recall from the lecture on Fractions that the key to understanding the cycle length in decimal notation was the prime factors of the numbers 1, 11, 111, 1111 etc. The same applies here but of course the numbers 1, 11 etc. are the base 12 forms. The first few are shown in the table above.

Remember that in order to find the cycle length for say 5, we look in the table for the first number that has 5 as a factor. The first one is 1111_{12} (1885 in decimal) and so the cycle length for 5 is 4. Similarly we see that 1/13 has a cycle of length 2, and that 1/7 has a cycle of length 6.

Perhaps amazingly this shows that 1/157 has a cycle of length 3 and even more strange that 1/22621 has a cycle of length 5. Well, it is not that strange, since

$1/157 = 1/111_{12} = 0.00e00e00e...$
$1/22621 = 1/11111_{12} = 0.0000e0000e...$

What is perhaps strange is that 157 and 22621 are prime. In the corresponding table for decimals, the first prime (apart from 11) was 1111111111111111111.

An important base is binary which is fundamental to the digital computer. We can similarly find the forms of recurring fractions in binary. For example

$1/3_{10} = 0.01010101...$

To prove this we write (using binary throughout)

$x = 0.01010101...$
$100x = 1.01010101...$ and then subtracting
$11x = 1$
$x = 1/11_2 = 1/3_{10}$

Another example of a recurring fraction is

$1/5 = 0.00110011...$

	base 2	decimal	factors	in base 2
1	1			
2	11	3	3	11
3	111	7	7	111
4	1111	15	3.5	11.101
5	11111	31	31	11111
6	111111	63	$3^2.7$	$11^{10}.111$
7	1111111	127	127	1111111
8	11111111	255	3.5.17	11.101.1001
9	111111111	511	7.73	111.1001001
10	1111111111	1023	3.11.31	11.1011.11111
11	11111111111	2047	23.89	10111.1011001
12	111111111111	4095	$3^2.5.7.13$	$11^{10}.101.111.1101$
13	1111111111111	8191	8191	1111111111111
14	11111111111111	16383	3.43.127	11.101011.1111111

Factors of 11, 111, 1111 etc in base 2.

which has a cycle length of 4. Again remember that the cycle length for a prime fraction $1/p$ is always a factor of $p-1$. Moreover, we find

$$1/7 = 0.001001...$$

which has a cycle length of 3 whereas in both decimal and in base 12 the cycle length is 6.

The cycle length follows the same rule as for other bases. The factors of 1, 11, 111 etc. in base 2 are shown in the table above. To find the cycle length for $1/p$ where p is prime we scan down the table of factors for the first entry with p. In the case of 5 we find that 1111 in binary (that is 15 in decimal) is the first with a factor of 5 so the cycle length for 1/5 is 4.

Similarly we see that that 1/7 has a cycle length of 3 and that 1/11 has a cycle length of 10.

The numbers 1, 11, 111 etc. in base 2 are of course the Mersenne numbers 2^n-1 discussed in the lecture on Amicable Numbers. In the case of the Mersenne primes, 1/31 has cycle length of 5, 1/127 has a cycle length of 7, and 1/8191 has a cycle length of 13. Indeed, if a Mersenne number M_n is in fact a prime then the cycle length of $1/M_n$ is clearly just n from the rule.

One would intuitively expect that the cycle length in binary would normally be longer than in decimal. But that is not generally the case as is illustrated by the table overleaf which gives the lengths for the first few primes.

Now we know that the number of digits in a cycle for a recurring binary fraction for $1/p$ cannot exceed $p-1$, and we know that the cycle length is given by the first of the numbers in the table for the factors of 11, 111 etc. having a

prime	in binary	in decimal	prime	in binary	in decimal
3	2	1	47	23	46
5	4	0	53	52	13
7	3	6	59	58	58
11	10	2	61	60	60
13	12	6	67	66	33
17	8	16	71	35	35
19	18	18	73	9	8
23	11	22	79	39	13
29	28	28	83	82	41
31	5	15	89	11	44
37	36	3	97	48	96
41	20	5	101	100	4
43	14	21			

Cycle lengths of prime fractions in binary and decimal up to 101.

factor of p. It follows that every prime number p must be a factor of the Mersenne number $M_{p-1} = 2^{p-1}-1$.

For example, 5 must be a factor of $M_4 = 15$ which it is; 7 must be a factor of $M_6 = 63$ (it is); 11 must be a factor of $M_{10} = 1023$ (it is) and 13 must be a factor of $M_{12} = 4095$ (it is).

This doesn't apply just to binary numbers of course; applying the argument to any base b we find that any prime number must be a factor of b^n-1 for some n. Indeed, since the cycle length must be a factor of $p-1$ it follows that any prime p must be a factor of $b^{p-1}-1$ for any b. There is a proviso: we have deduced this by considering recurring fractions. But if the base b has p as a factor then it won't recur and so the argument fails.

Let's try this out on that favourite prime 7. We have just seen that 7 is a factor of $2^6-1 = 63 \doteq 7 \times 9$. We also find

$3^6-1 = 728 = 7 \times 104$
$4^6-1 = 4095 = 7 \times 585$
$5^6-1 = 15624 = 7 \times 2232$
$6^6-1 = 46655 = 7 \times 6665$
$7^6-1 = 117648 = 7 \times 16806 + 6$ fails with base 7
$8^6-1 = 262143 = 7 \times 37449$
$9^6-1 = 531440 = 7 \times 75920$
$10^6-1 = 999999 = 7 \times 142857$

So, as anticipated, it fails with base 7 because 1/7 in base 7 does not recur and is simply 0.1.

We can also look at the situation from the other point of view; that is consider the various primes for a fixed base. We might as well use base 10.

We find

$10^2 - 1 = 99 = 3 \times 33$
$10^4 - 1 = 9999 = 5 \times 1999 + 4$ fails with prime 5
$10^6 - 1 = 999999 = 7 \times 142857$
$10^{10} - 1 = 9999999999 = 11 \times 909090909$

So it fails with the prime 5 since 5 is a factor of the base 10 and 1/5 does not recur as a fraction in base 10. But it works for 3, 7, and 11 which are relatively prime to 10.

By moving the minus one to the other side of the equation, we can rearrange the fact that has emerged to say that if p is prime and b is not a multiple of p then

$b^{p-1} \equiv 1 \bmod p$

This is Fermat's Little Theorem which he mentioned around 1640 but gave no proof. This was typical of Fermat. Euler gave a proof of the Little Theorem in 1736.

We have encountered this theorem almost by accident by a very unorthodox approach concerning recurring fractions. We will now look at it using a more traditional and indeed straightforward description.

Fermat's Little Theorem

JUST TO RECAP, the theorem says that if p is any prime number and some other number a is not a multiple of p, then a raised to the power of $p-1$ is congruent to 1 modulo p. That is

$a^{p-1} \equiv 1 \bmod p$

For example, if $p = 7$ and a is 8 then $8^6 = 262144$ has remainder 1 when divided by 7. And indeed $262144 = 37449 \times 7 + 1$. This was one of the examples given above. Similarly if p is 11 and a is 2 then we find that $2^{10} = 1024 = 93 \times 11 + 1$. It almost seems like magic.

The conditions are important. For example if p is 9 (and so not prime) and a is 2 then $2^8 = 256$. But 256 divided by 9 has remainder 4 so it fails.

Similarly, if p is prime but a is a multiple of p, then clearly a^{p-1} is exactly divisible by p. Thus if p is 3 and a is 6 then $6^3 = 216$ is obviously divisible by 3 and so the remainder is zero.

Here is a traditional proof supported by an illustrative example; we examine the case when p is 5 and a is 8. Consider the multiples of a thus

$a, 2a, 3a, 4a, ..., (p-1)a$

For illustration with $p = 5$ and $a = 8$ these are

$$8, 16, 24, 32$$

Now these numbers are not congruent to each other mod p, nor is any one congruent to zero. Therefore they must be congruent to the set $1, 2, ..., p-1$ in some order.

Indeed in the example we find they are

$$8 \equiv 3 \bmod 5, \quad 16 \equiv 1 \bmod 5, \quad 24 \equiv 4 \bmod 5, \quad 32 \equiv 2 \bmod 5$$

Now multiply all these congruences together thus

$$a^{p-1} \times (p-1)! \equiv (p-1)! \bmod p$$

and now simply cancel the $(p-1)!$ and get

$$a^{p-1} \equiv 1 \bmod p$$

as required. Cancellation is allowed because the number being cancelled is relatively prime to the modulus as was mentioned in the lecture on Amicable Numbers. And in this case $(p-1)!$ is relatively prime to p so all is well.

In the example we have

$$8 \times 16 \times 24 \times 32 \equiv 3 \times 1 \times 4 \times 2 \bmod 5$$

and so cancelling $1 \times 2 \times 3 \times 4$ on both sides we get

$$8^4 \equiv 1 \bmod 5$$

and that's it.

The theorem provides a useful shortcut for some calculations. Suppose we wish to evaluate 5^{42} mod 11. Now taking $b = 5$ and $p = 11$, the little theorem tells us that $5^{10} \equiv 1 \bmod 11$. So

$$5^{42} \bmod 11 = (5^{10})^4 \times 5^2 \bmod 11 \equiv 1 \times 25 \bmod 11 \equiv 3 \bmod 11$$

Interestingly, the converse of the Little Theorem does not hold. In other words if $b^{p-1} \equiv 1 \bmod p$ for all b relatively prime to p, then it does not follow that p is prime. Thus we find

$$a^{560} \equiv 1 \bmod 561$$

for all a that are relatively prime to 561 yet 561 is $3 \times 11 \times 17$. Such numbers are called Carmichael numbers after the American mathematician Robert Carmichael (1879–1967).

Further reading

THE BABYLONIAN number system and the introduction of zero are discussed in *The Nothing That Is* by Robert Kaplan.

Modular arithmetic is discussed in any book on elementary number theory. For example *Elementary Number Theory* by David Burton and *Elementary Number Theory and its Applications* by Kenneth Rosen give good coverage. A classic work is *The Higher Arithmetic* by H Davenport – the seventh edition has a final chapter relevant to the computer age written by his son J H Davenport.

On a lighter note, *Amusements in Mathematics* by Henry E Dudeney has many numerical puzzles based on concepts such as digital roots.

Volume 12 of *Contemporary Mathematics* by Brillhart, Lehmer, Selfridge, Tuckerman and Wagstaff contains tables giving factorizations of $b^n \pm 1$ for various values of the base b.

As an example of the use of tests for divisibility see Appendix D on Polydivisibility.

Exercises

1. What is the value of ⟨⟨⟨𝖸 ⟨⟨⟨𝖸 ? Assume there is no space at the end. Write the answer in Egyptian and Roman numbers as well as in normal decimal notation.

2. What is 3/4 as a hexadecimal fraction? As an octal fraction?

   What are 1/4, 1/2 and 3/4 as base 7 recurring fractions? Similarly what are 1/5, 2/5, 3/5 and 4/5 as base 7 recurring fractions?

3* Find the factors of 11, 111, ..., 11111111 expressed in bases 7 and 8. Present them in the style of the table for base 12.

7 Bells

THIS LECTURE looks at bell ringing (the posh name is Tintinnalogia) which as an organized activity has been around for a long time. Indeed, an important book by Stedman on the subject was published in 1668 (two years after the Great Fire of London). As well as illustrating various sequences of changes we also look at some of the mathematical ideas behind the sequences such as permutations and other simple aspects of group theory.

Rounds and plain hunting

WE NORMALLY think of a bell as hanging mouth down. If we swing the bell, then it will strike at regular intervals determined by its natural frequency of swinging. Like a pendulum, this does not vary much with the size of the swing but varies with the size of the bell.

If a tower has several bells and they are tuned to say C, E, G, C', then however they are sounded they will be harmonious since the major common chord CEGC' is harmonious.

Moreover, they will be repeated at different intervals because the rate of swinging will be different. If they are not in harmony or we add more bells and continue to swing them mouth down then the result will be a cacophonic mess.

In English bell towers, the bells are sounded with the mouth nearly upwards. The bell swings nearly through 360 degrees with strikes at each end of the swing in either direction. As a consequence, the ringer is able to hold or hasten the bell after each strike and so control the repetition of the strikes.

Note that English practice is not to use bells as just another musical instrument on which melodies can be played or to play multipart chords but to explore the elegance of varied sequences of single notes.

So the tradition has arisen that bells sound best when they are rung in sequence with a uniform gap between each sound. The basic sequence is so-called rounds in which the bells are rung in order over and over again starting with the highest and ending with the lowest. If there were four bells with notes C, D, E, F, then the order of ringing rounds would be FEDCFEDCFEDC....

We usually number the bells with 1 being that with the highest note. So the rounds with four bells are written in the abstract form 123412341234.... It is conventional to put a space at the end of each row thus 1234 1234 1234 .... For technical reasons alternate rows are a little different. This is because bells strike twice on each cycle of swings. One is called the handstroke and the other the backstroke. There is a pause after every other row equal to the time of one strike. This pause makes it easier to understand the pattern of ringing.

Nice Numbers

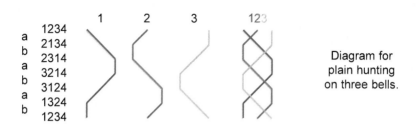

a	1234
b	2134
a	2314
b	3214
a	3124
b	1324
	1234

Diagram for plain hunting on three bells.

Just ringing rounds is a bit boring so the idea of making a change after each row was introduced. The simplest change is where two bells change places. Thus 1234 might be followed by 1324 in which the middle two bells have changed places. It is important that a bell never move by more than one place since this would be too hard for the ringer. Moreover, the heaviest bell (in this example 4) is often left in the same place each time because it is harder to change the position of the heaviest bell. This also accentuates the rhythm.

With this requirement there are only 6 ways (= 3!) to ring the four bells. The goal is to ring these 6 rows without repetition and only with adjacent bells changing places. There are only two ways of doing this and one is the reverse of the other. Thus we have

1234 2134 2314 3214 3124 1324 1234

1234 1324 3124 3214 2314 2134 1234

Note that with 4 bells (and keeping the last fixed) there are only two moves that can be made at any point. Either we interchange the first two bells or we interchange the middle pair. One change will take us back to the previous row so there is only one way forward. So having decided on the first change the rest follow without choice.

Changes are often depicted by a zigzag diagram as shown above. Individual bells can be shown separately or superimposed. Different colours are used for clarity. (The path of a bell is often known as a blue line.) The letters on the left identify the change being made. Change *a* interchanges the first two bells and change *b* interchanges the middle two bells.

Note carefully that each bell follows the same pattern: in the same place for one row, then shifting left for two rows, the same place for one more row, and then shifting right for two rows. But they differ in that they start at different points in the cycle rather like singing rounds. This very simple pattern is called hunting.

With more bells there are many more combinations and more ways to ring them all. Thus with 8 bells (a common number – an octave) and the last fixed there are 5040 (= 7!) possible rows. To ring all these typically takes about three hours and is known as ringing a full peal. Bell towers often have boards listing events when full peals were rung.

With 5 bells (and keeping the last fixed) there are 24 (= 4!) possible rows. Moreover, it is possible to interchange two pairs of bells in the same change. Thus

12345   21435   24135   42315   and so on

The first change interchanges two pairs of bells. Such changes are known as doubles. Similarly, with 7 bells (and again keeping the last fixed), we can interchange three pairs of bells. Such changes are known as triples.

In mathematics, the different arrangements of say 123 are known as permutations. Permutations are either even or odd according as an even or odd number of single interchanges are required to give the permutation starting from 123. Alternate permutations in

123   213   231   321   312   132   123

are thus even and odd because each is obtained from its predecessor by a single interchange.

Note carefully that if we have enough bells so that we can do doubles, then we cannot go through all possible rows by just using doubles because that would only cover all the even permutations but not the odd ones.

Let us return to three bells (the fourth bell being fixed) and consider the mathematics a bit more. There are two possible independent operations

a (1, 2)   interchange first two

b (2, 3)   interchange middle pair

We write (1, 2) meaning the bell in position 1 goes to position 2 and 2 goes to position 1. In general permutation theory (1, 2, 3) would mean that the bell in position 1 goes to 2, that in position 2 goes to 3, and that in position 3 goes to 1. But we don't allow that in bell ringing because the bell in position 3 would move two places. Similarly we are not allowed (1, 3). However that could be achieved by doing a, then b, and then a again – written aba. So it is not independent. Moreover, the same could be achieved by doing bab.

Note carefully that ab (that is do a and then b) is not the same as ba. The operations do not commute. This is not unusual, the operations on a square of rotating it by a right angle and flipping it about the vertical axis do not commute either – the result depends upon the order in which they are done as shown below.

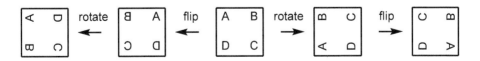

Rotation and flipping of a square do not commute. Starting from the centre, flip then rotate gives far left square, rotate then flip gives far right one.

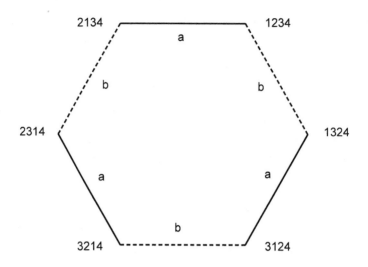

Graph showing the changes on three bells.

We can do various other interesting bits of algebra with these operations. For example note that doing *a* and then *a* again has no effect. We say that $a^2 = 1$. Similarly $b^2 = 1$. Furthermore, the whole sequence of six changes is *ababab* or $(ab)^3$ and this returns us to the starting point. So $(ab)^3 = 1$.

However, we have seen that *bab* = *aba*, so

ababab = *a*(*bab*)*ab*

= *a*(*aba*)*ab* = *aabaab*

Now $a^2 = 1$ and $b^2 = 1$, so finally we get

$(ab)^3$ = *ababab* = *aabaab* = $1.b.1.b$ = *bb* = 1

which confirms that *ababab* gets us back to the beginning.

Another way of depicting the sequence is by drawing a graph with points representing each row and lines between them identifying the changes. In this case we get a hexagon as shown above. The lines for the change *a* are shown solid whereas those for *b* are dashed.

If we have four bells moving and a fifth fixed then the possibilities are much greater. The basic moves are

a	(1, 2)(3, 4)	interchange two pairs – a double
b	(2, 3)	interchange 2 and 3
c	(3, 4)	interchange 3 and 4
d	(1, 2)	interchange 1 and 2.

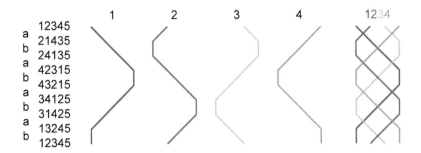

a	12345
b	21435
a	24135
b	42315
a	43215
b	34125
a	31425
b	13245
	12345

Alternate doubles and singles on four bells.

As it happens we will not use *d* for the moment. In any event it is equivalent to *ac* (or *ca*).

If we alternately do *a* and *b* then we get the situation shown above. The bells again do plain hunting. Note how the paths of bells 2 and 3 are mirror images as are the paths of bells 1 and 4. But this sequence alone is no good because after 8 changes we are back to the beginning whereas we know that we have to do 24 changes for four moving bells in order to cover all possibilities.

The solution is to introduce another change and thus avoid the return to rounds. This change breaks the plain hunting.

Plain Bob Minimus

THE SOLUTION is to do change *c* rather than change *b* after 7 changes. This gets us into a different cycle of 8 and then we do another *c* to prevent that repeating and a final *c* to finish off. The result then covers all 24 changes. We can write the whole sequence as

$((ab)^3 ac)^3$

This sequence is known as Plain Bob Minimus and is shown overleaf. Minimus simply refers to the fact that there are 4 moving bells. Other terms are Doubles (5 bells), Minor (6), Triples (7), Major (8), Caters (9), Royal (10), Cinques (11), and Maximus (12).

Note that bell 1 (the highest or treble) continues to do plain hunting, whereas the paths of the three other bells have kinks in them. These kinks due to the introduction of the change *c* are known as the work and the bells are said to be dodging because they dodge back and forth rather than continue smooth hunting. Note also that the paths of the working bells 2, 3, and 4 are again identical but simply shifted in time.

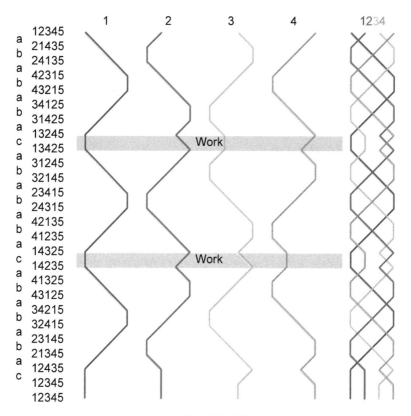

Plain Bob Minimus.

Furthermore, within each group of eight changes, the bells again have paths that are mirror images in pairs. We saw that in the first group, the pairs were numbers 1, 4 and 2, 3. In the second group they are 1, 2 and 3, 4; and in the third group they are 1, 3 and 2, 4.

Another way to ring the changes on four bells (with a fixed fifth bell) uses Single Changes only. This style of ringing was used in earlier times when the bearings of the bells were less smooth and making changes was consequently somewhat more difficult. The changes used are those called *b*, *c*, and *d* above. The result is shown opposite. This method is known as Double Canterbury.

Bell 1 again does plain hunting. The other three have a more languid pattern whereby a bell stays in the same place for three rows. However, they do all have exactly the same sequence although shifted in time. Altogether, it doesn't look so pretty and it also sounds a bit monotonous.

The mathematical sequence is not quite so clear either. Basically, it consists of alternate groups *cdc* and *dcd* joined by *b* and repeated three times thus

$$((cdc)b(dcd)b)^3$$

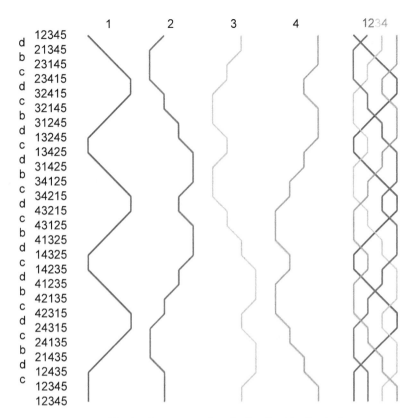

Single changes on four bells: Double Canterbury.

but this is not very clear because it starts part way through one of the sequences *dcd*.

The next question we might ask is can we represent these possibilities with four moving bells as a graph like the hexagon for three moving bells. The answer is yes but it gets complicated.

The graph overleaf shows the 24 possible rows (bell 5 is omitted for simplicity since it is fixed) and the allowed changes joining them. Different styles of lines identify the different kinds of changes. Thus solid lines represent the doubles. Note carefully that the opposite sides of the graph have to be identified. So the lines coming out of the point marked A at the bottom right are the same lines coming out of the corresponding point A at the top left.

Topologically, the figure consists of a map of six squares, four hexagons, and three octagons drawn on a real projective plane. Each point is where a square, hexagon, and octagon meet.

In order to construct a set of changes all we have to do is to trace a path through all 24 points that visits them all just once. This can be done in several ways.

The two methods we have encountered, Plain Bob Minimus and Double Canterbury, are shown on the next two graphs. In each case the path actually taken is shown in bold red and blue in the direction shown by the arrows.

In the case of Plain Bob Minimus shown opposite every double change is used. There are actually three octagons in the graph. One is in the centre and the other two are split – one across the sides and one from top to bottom.

The sequence starts from 1234 (in bold) by going around the central octagon (in red) with alternate doubles (solid line) and middle singles (dash dot) (this is $(ab)^3a$) and then changes on a blue dotted line (c, which is the work) from 1324 to 1342 to go around the octagon across the sides and then changes on another

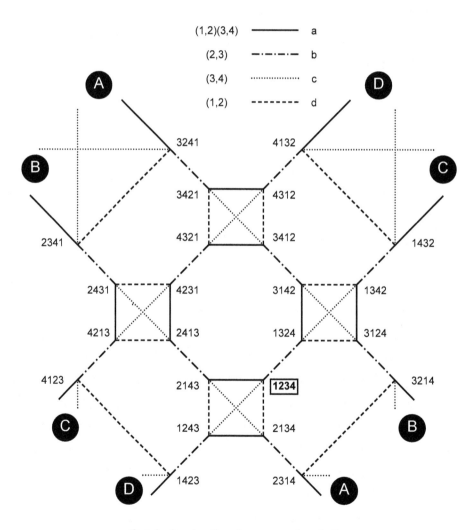

Graph showing the changes on four bells.

blue dotted line (c) from 1432 to 1423 to go around the third octagon and finally taking another blue dotted line (c) from 1243 to 1234 to finish. Hence we get

$$((ab)^3 ac)^3$$

By contrast, the graph overleaf shows the sequence using Double Canterbury. It is quite different. No doubles (a) are used and so none of the solid lines is traversed. But many of the dashed lines (d) are used instead.

The basic theme is that the path consists of excursions around the four points of the six squares using dashed and dotted lines (in red). These correspond to

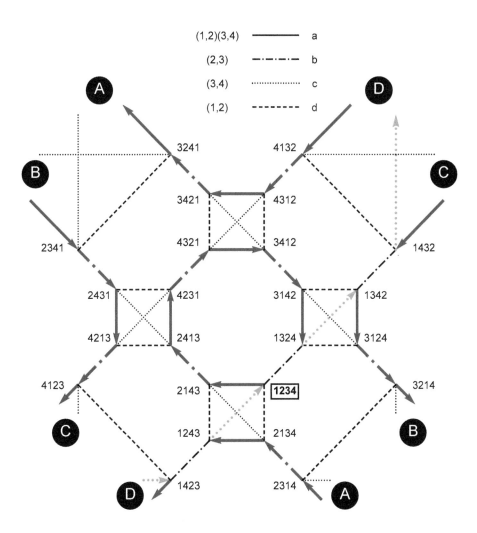

The path of Plain Bob Minimus.

146 Nice Numbers

changes (1, 2) on the first pair and changes (3, 4) on the last pair. In some cases the path uses both diagonals (*cdc*) and in others it only uses one (*dcd*). And these squares are then joined by the blue dash-dotted lines (*b*) which represent the changes on the middle pair (2, 3).

Hence we see how we get

$$((cdc)b(dcd)b)^3$$

So we have looked at this graph as either basically three octagons or as six squares joined together. In Plain Bob Minimus, we go around the octagons and with single changes we go around the squares.

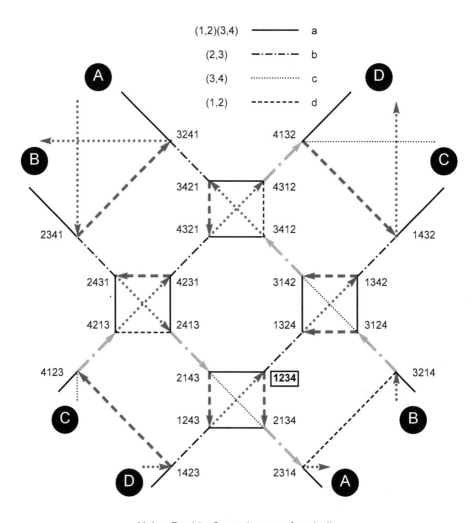

Using Double Canterbury on four bells.

However, we can also look upon the graph as comprising four hexagons. Each hexagon consists of those changes where one of the four bells is fixed at the back. This provides another way of ringing all the changes. Go around each hexagon in turn and then jump between them as necessary. The method known as Single Court Minimus is almost like this and is discussed as one of the Exercises (but the treble always does plain hunting).

This is an interesting way of looking at the graph because it shows how the graph for four bells relates to that for three bells which was just a hexagon. This also hints at how the graph could be extended to five moving bells. We just replicate the graph for four bells five times and join up the points as necessary. Clearly it gets too complex to contemplate!

Plain Bob Doubles

THE PLAIN BOB METHOD can also be rung on five moving bells and is known as Plain Bob Doubles. We saw how 8 changes on four bells gave plain hunting and returned to rounds. Similarly 10 changes on five bells do the same. We need two doubles

      a   (1, 2)(3, 4)   interchange first two pairs

      b   (2, 3)(4, 5)   interchange last two pairs

Again we see that every bell does plain hunting but they are staggered.
In order to do the complete extent we also need two single changes

      c   (3, 4)   interchange 3 and 4

      d   (2, 3)   interchange 2 and 3

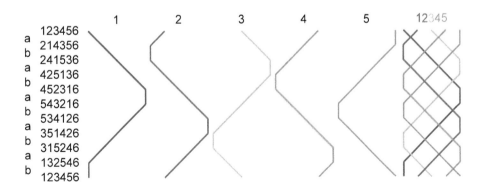

Plain hunting on five bells.

In order to prevent the return to rounds we do change c rather than b at the end so that the sequence becomes

$(ab)^4 ac$

which should be compared with the sequence $(ab)^3 ac$ of Plain Bob Minimus. We then carry on plain hunting for another 9 changes and then do c again. If we do this group of 10 changes four times giving a total of 40 changes thus

$((ab)^4 ac)^4$

then this will return us to plain hunting once more.

Note the analogy with Plain Bob Minimus which in total was $((ab)^3 ac)^3$. The first 20 or so are shown below. Note how the bell in place 5 at each point of work (when the treble leads) does four strikes in the same place. This is a characteristic of the method on an odd number of moving bells. This sequence of 40 changes is known as a Plain Course.

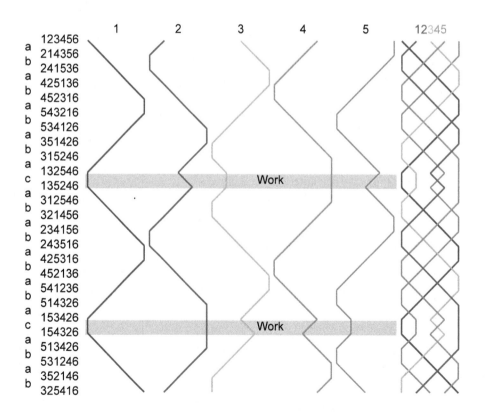

The first 24 changes of Plain Bob Doubles.

However, we need 120 changes altogether and so a further modification is required to prevent it reverting to rounds after 40 changes. In fact the change *c* is replaced by the change *d*. This interchanges 2 and 3 rather than 3 and 4. The net effect is that the bells in positions 2, 3, and 4 are cycled around. This manoeuvre is known as a Bob.

The diagram below shows changes 21 to 44. The change *d* keeps the bells in places 1, 3, and 5 unmoved. Again the bell in place 5 does four consecutive strikes in that place. But exciting dodging is not involved. All that really happens is that one bell reverses its hunting early.

Change *d* (a Bob) is also made after 80 changes and finally at the end after 120 changes which then returns immediately to rounds. So the complete sequence for Plain Bob Doubles is

$$(((ab)^4ac)^3(ab)^4ad)^3$$

The final end of Plain Bob Doubles is shown overleaf. The last Bob returns immediately to rounds.

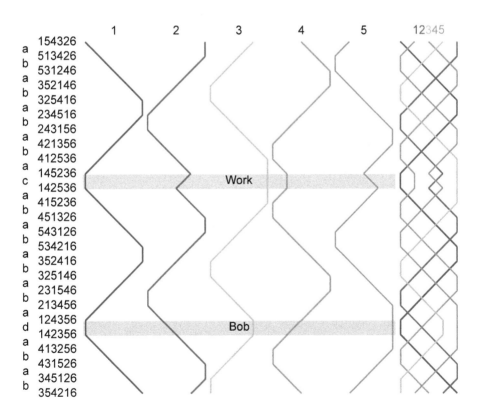

Changes 21 to 44 of Plain Bob Doubles.

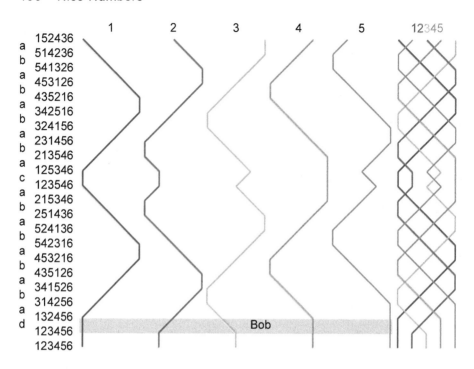

The final end of Plain Bob Doubles.

Ringers know the plain course but a caller will indicate when a Bob is to be done by calling "Bob" and then when it is about to return to rounds by calling "That's all". Note that the full extent does not have to be rung exactly as indicated. A Bob can be called at any time (but inevitably when the treble is at the front). For example if Bob were called at change 30 and then again at 70 and 110 then we would find that we had to do 10 more changes before returning to rounds. There are in fact three sets of 40 changes – the basic Plain Course and two variations obtained by rotating bells in positions 2, 3, and 4. We can swap between these three sets by doing Bobs at appropriate times and so can ring the full extent in many different ways.

Note carefully that the term Bob refers to the change d. The method Plain Bob Minimus with only four moving bells does not actually include any Bobs!

The same approach can be used with more bells. An additional kind of interchange known as a Single will be required to prevent premature returning to rounds if a full extent is to be rung. Thus with 6 moving bells we have

- a (1, 2)(3, 4)(5, 6)   interchange three pairs – a triple
- b (2, 3)(4, 5)   interchange middle two pairs
- c (3, 4)(5, 6)   interchange last two pairs

So a Plain Course of Plain Bob Minor is

$((ab)^5 ac)^5$

which is similar to that for Plain Bob Doubles with 4 replaced by 5. Thus it has 60 changes. In order to do a full extent we also need

    d $(2, 3)(5, 6)$ the Bob
    e $(5, 6)$ the Single

and by calling these at appropriate points we can do the full extent of 720 changes.

Stedman

FABIAN STEDMAN (1640–1713) wrote two famous books on bell ringing, *Tintinnalogia* in 1668 and *Campanalogia* in 1677. He introduced a number of new methods for ringing bells and his name is immortalized by the names of these methods. The title page of the third edition of Campanalogia (dated 1733 and thus posthumous) says

> By Plain and Methodical Rules and Directions, whereby the Ingenious Practitioner may, with a little Practice and Care, attain to the Knowledge of Ringing all Manner of *Double*, *Tripple*, and *Quadruple Changes*.
>
> With Variety of *New Peals* upon Five, Six, Seven, Eight, and Nine Bells. As also the Method of calling *Bobs* for any *Peal* of *Tripples* from 168 to 2520 (being the *Half Peal*:) Also for any *Peal* of *Quadruples*, or *Cators* from 324 to 1140.

One idea of Stedman's was to break the bells up into groups. For example, with five moving bells, we break the bells into a group of three and a group of two. Suppose the group of three do the changes on three bells discussed earlier whereas the group of two just keep dodging. After six changes this would return to rounds as shown below.

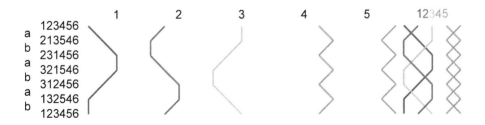

152 Nice Numbers

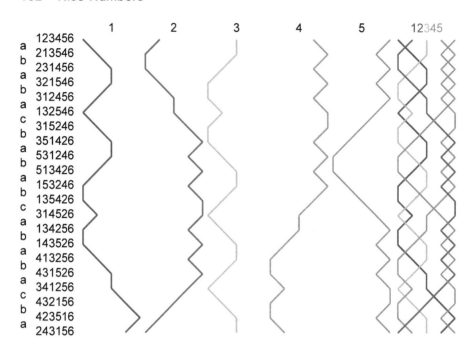

The first 20 changes of (our variation of) Stedman Doubles.

In order to avoid this we exchange one bell between the two groups. The changes are

a (1, 2)(4, 5)

b (2, 3)(4, 5)

c (1, 2)(3, 4)

and so basically we have *ababa* followed by *c* and then *babab* followed by *c*, and so on. Note that alternate groups of changes on the threes are reversed in direction. If this were not done the first two bells would dodge too much. The first few changes are shown above. Note carefully that this is not as Stedman is normally started but simplifies the mathematics of the permutations.

Although it might not be obvious, in fact every bell follows exactly the same pattern but they are phased differently. Thus bell 2 repeats the pattern of bell 1 but 12 changes later and bell 5 follows 12 changes after that and so on. The individual patterns look complicated although the superimposed pattern has an elegant symmetry.

After 60 changes we have done a plain course

$$((ab)^2 ac(ba)^2 bc)^5$$

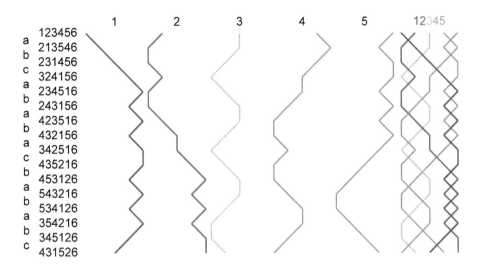

The proper start of Stedman Doubles.

and we return to rounds. However, this means that we have only done half of the total extent of 120. The reason is obvious. The changes *a*, *b*, and *c* are all doubles and so we have only dealt with the even permutations. In order to do the other 60 we must insert a single (twice). We can use

$$d \quad (2, 3) \qquad \text{the Single}$$

and if we use it at the end the full extent will be

$$(((ab)^2 ac(ba)^2 bc)^4 (ab)^2 ac(ba)^2 bd)^2$$

Note some interesting patterns. In particular, note how bell 5 goes from position 5 to 1, does a double strike on 1, and then goes out to 5 again. This is called running in quick and running out quick respectively. In practice bell ringers do not start quite as we have done but with the treble running out quick. This means that they start part way through a sequence *babab* and this somewhat spoils the mathematical analysis (though maybe sounds prettier). This proper start is shown above. (Grammarians would prefer to run in and run out quickly no doubt.)

One consequence of starting part way through a touch of six changes thus

abcababac...

is that the Single *d* is also applied part way through a touch thus

...bacbadabcab...

154 Nice Numbers

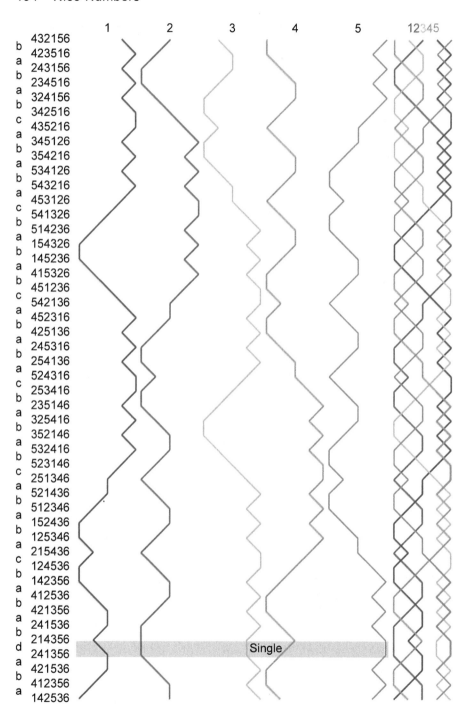

Changes 19 to 63 of (our variation of) Stedman Doubles showing the Single.

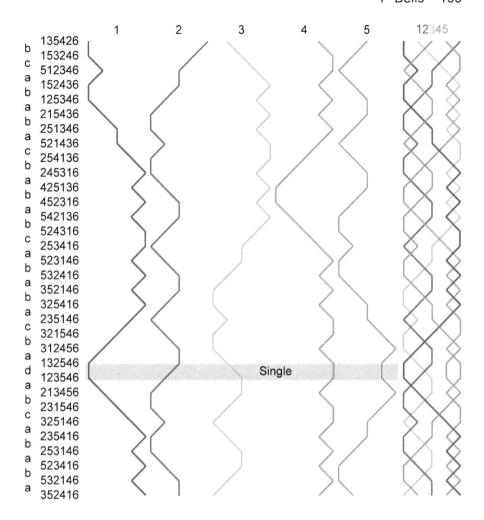

Part of normal Stedman Doubles showing the Single.

and as a consequence the full sequence as normally rung is

$$(abc((ab)^2ac(ba)^2bc)^4(ab)^2acbad)^2$$

The corresponding part of the normal Stedman including the Single is shown above.

Note that we cannot really do Stedman with less than five moving bells since we need at least three plus two.

Indeed, Stedman really only works on an odd number of moving bells. So we can have seven bells with one group of three and two groups of two; the bells then change between the various groups.

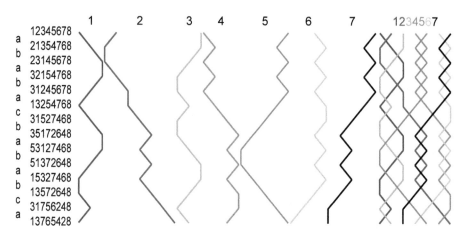

The first few changes of (our version of) Stedman Triples.

This is known as Stedman Triples and the key changes are

    a (1, 2)(4, 5)(6, 7)

    b (2, 3)(4, 5)(6, 7)

    c (1, 2)(3, 4)(5, 6)

The basic course is as in Stedman Doubles namely $((ab)^2ac(ba)^2bc)$. Whereas in Doubles this reverts to rounds if applied 5 times, in Triples it reverts to rounds if applied 7 times (after 84 changes). Again, as normally rung, it starts with the treble running out quick and so it starts $abc(ab)^2ac$....

In order to do a full extent we can use a Bob and a Single. These are

    d (1, 2)(3, 4)(6, 7)   the Bob

    e (1, 2)(3, 4)   the Single

Unlike Stedman Doubles, the Bob and Single simply replace an instance of *c* and thus we have sequences as follows

    cababadbababc

    cababaebababc

Curiously, the Single is in fact a double! Moreover, the normal changes and the Bob are all triples and so change the permutation from odd to even and back at each change. Thus, unlike Stedman Doubles, we do not get locked into even permutations and so do not inevitably have to use the Single to do the odd permutations. It has been shown as recently as 1994 that it is indeed possible to ring a full extent of Stedman Triples using only the changes *a*, *b*, *c*, and *d*.

Grandsire

NO OVERVIEW of bell ringing would be complete without a mention of the Grandsire methods. In the plain Bob methods one bell does plain hunting. In Stedman's, no bells do plain hunting. But in Grandsire methods two bells do plain hunting (more or less). Below is shown the start of Grandsire Doubles.

It is bells 1 and 2 that do plain hunting but they cannot be adjacent since otherwise no bell could pass them because it would have to move two places and that is not allowed. So we start with a change that avoids this. We have

$\quad a \quad (1, 2)(3, 4)$

$\quad b \quad (2, 3)(4, 5)$

$\quad c \quad (1, 2)(4, 5)$

and the sequence basically is simply

$\quad cb(ab)^4$

repeated.

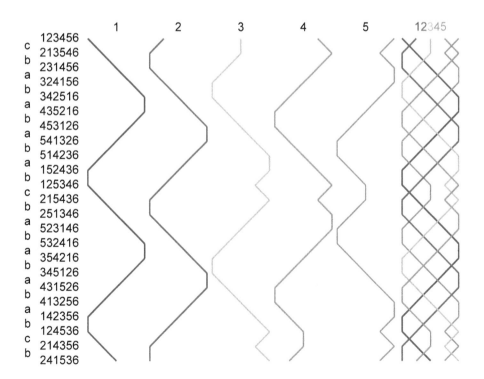

The first 22 changes of Grandsire Doubles.

158 Nice Numbers

Note that bells 3, 4, and 5 follow a pattern almost identical to the working bells in Plain Bob Minimus.

After 30 changes it returns to rounds. In order to do a full extent (120) two important points must be noticed. First, the changes used so far are all doubles and so we must do a single to do the odd as well as the even permutations. Moreover, if bell 2 always follows bell 1 two places behind, then it cannot possibly cover the full extent because that requires bell 2 to be in all possible places after bell 1. The solution is to arrange for the other three bells to take their turn at doing plain hunting two places behind bell 1.

We can achieve this by introducing a single, namely

    d (4, 5)   the Single

In fact we replace the b just before the change c by the single d. This has the effect of reversing the roles of bells 2 and 3. So bell 3 now does plain hunting and follows bell 1 as bell 2 previously did. But this is not enough since if we do it again at 60, all that happens is that bells 2 and 3 swap back and so we would revert to rounds. To avoid this we also replace the previous change a by c so that the sequence becomes ...*abcdcba*.... The net result is that 3 now does plain hunting and 2, 4, and 5 are cycled around as well as shown below.

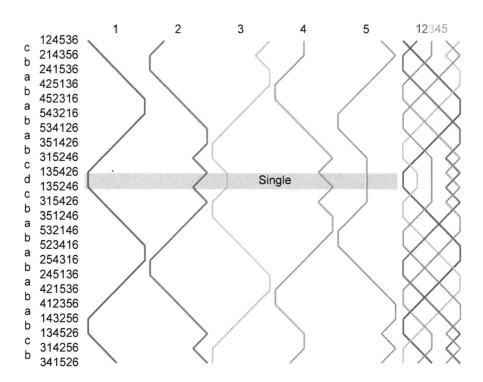

Changes 20 to 42 of Grandsire Doubles showing a Single (d).

So the initial plain round 12345 is replaced by 13524. At 60 it is replaced by 15432 and at 90 by 14253. And then at 120 it reverts to rounds having done the full extent.

The replacement of 12345 by 13524 is the operation (2, 3, 5, 4). Since this has length 4, it follows that applying it four times returns to the original state as we desire.

A Bob is also defined for Grandsire Doubles. However, this does not involve an additional permutation. It simply consists of replacing the normal *ababcba* by *abcbcba*. In other words we simply replace the *a* which is two before the *c* by another *c*. This replaces 12345 by 15324. This has the same effect as the Single except that 3 and 5 are interchanged. So 5 is now the bell that does plain hunting and follows 1. So it is the operation (2, 5, 4) and bell 3 is unaffected. It is not very useful because it still leaves us with even permutations.

Groups

THE MATHEMATICS behind all this is called group theory. A group is a set of things which have operations on them. In our case they are the various permutations of *N* things and the operations are changes such as (2, 3). Groups can be subdivided into smaller parts some of which are genuine groups in their own right. Other important subdivisions are called cosets. See Appendix E for an introduction to some simple properties of groups.

In the case of Plain Bob Doubles, the Plain Course visits just a subgroup of size 40 of the whole group of permutations on 5 objects (which has size 120). The other two subsets of 40 are cosets (they are not groups in their own right since they do not have a plain round). Whenever we do a Bob we switch from one coset to another.

Further reading

CHAPTER 7 of *Music and Mathematics* edited by Fauvel, Flood and Wilson covers bell ringing from a gently mathematical perspective. A traditional bell ringers' book is *Bell-ringing* by Ron Johnston. Diagrams of many methods will be found in *Diagrams* in the Jasper Snowdon Series.

Facsimiles of *Tintinnalogia* and *Campanalogia* were published by Kingsmead Reprints (1970) and Christopher Groome (1990) respectively. The latter was a limited edition.

For deep mathematical insight (and possibly a headache) consult various papers by Arthur T White such as *Ringing the Cosets* in the *American Mathematical Monthly* for 1987 (pp 721–746) and *Fabian Stedman: The First*

160 Nice Numbers

Group Theorist? in the *American Mathematical Monthly* for 1996 (pp 771–778). They are available at www.jstor.org.

The topology of the projective plane is described in *Gems of Geometry* by the author. Also see Appendix E on Groups.

Exercises

1. The following method is known as Canterbury Minimus

    1234 2143 2413 2431 4231 4213 4123 1432

    1342 3124 3214 3241 2341 2314 2134 1243

    1423 4132 4312 4321 3421 3412 3142 1324

    1234

    Draw its diagram nicely.

2. Using the notation that the permutations are

    | a | $(1, 2)(3, 4)$ | interchange two pairs – a double |
    | b | $(2, 3)$ | interchange 2 and 3 |
    | c | $(3, 4)$ | interchange 3 and 4 |
    | d | $(1, 2)$ | interchange 1 and 2. |

    give the algebraic sequence for Canterbury Minimus. That is the form in which Plain Bob Minimus is $((ab)^3ac)^3$.

3*. The method for Single Court Minimus starts

    1234 2134 2314 3241 3421 4312 4132 1432 1342

    In the Graph for changes on four bells shown earlier, Single Court Minimus goes around three of the hexagons in turn although it starts part way through the hexagon for bell 4. The fourth hexagon (for bell 1) is visited in stages and is not traversed in one lump. This enables bell 1 (the treble) to do plain hunting as normal. Bells 2, 3, and 4 follow the same pattern but staggered; this pattern includes a sequence of six strikes.

    Complete the method, draw its diagram and give the algebraic sequence.

8 Primes

IN THIS LECTURE we look at some of the techniques for testing for primes. We also introduce complex numbers in order to complete the number system and then consider the existence of primes in various algebraic structures such as the complex (or Gaussian) integers and polynomials.

However, we start by considering the Greatest Common Divisor and some intriguing properties of the Fibonacci numbers.

Greatest Common Divisor

AN IMPORTANT CONCEPT is that of the greatest common divisor (gcd) of two integers. It is defined as the largest integer that divides each exactly with no remainder. It is sometimes called the highest common factor (hcf). An old term is the greatest common measure (gcm).

As an example consider 15 and 55. The largest number that divides into both of these is of course 5. So gcd(15, 55) = 5.

If the two numbers a, b are factored into their primes thus

$a = p_1 p_2 p_3 ...$

$b = q_1 q_2 q_3 ...$

then the gcd is found by taking those primes that are common to a and b. Thus in the case of 15 and 55 we have

$15 = 3 \times 5$ and $55 = 5 \times 11$

and so the gcd is just 5 since that is the only common factor. Multiple primes have to be taken into account. Thus consider 252 and 120

$120 = 2^3 \times 3 \times 5$

$252 = 2^2 \times 3^2 \times 7$

The highest common power of 2 is 2^2 and the highest common power of 3 is just 3. And the other prime factors are not in common. So

gcd(120, 252) = $2^2 \times 3 = 4 \times 3 = 12$

If two numbers such as 15 and 44 have no common factors then the gcd is 1 and we say that the numbers are relatively prime or coprime as mentioned earlier. Note that in this example neither of the numbers is itself a prime number.

The gcd was known to Euclid and the best way to find the gcd of two numbers is to use Euclid's algorithm. This is described by Proposition 2 of Book VII of the Elements. As usual it is hard to understand because of the geometrical approach used by Euclid.

The basic principle of the algorithm is that we keep taking away multiples of one number from the other alternately until finally one exactly divides the previous one. When this happens the final divisor is the greatest common divisor or gcd.

Let's use this algorithm on the examples above. First take 15 and 55. We divide 55 by 15 which gives 3 and remainder 10. We now do the same with 10 and 15. 15 divided by 10 gives 1 and remainder 5. Then 10 divided by 5 gives 2 and remainder 0. When the remainder is zero, the last divisor, in this case 5, is the answer. This can be laid out as follows

```
15)55
   45
   ──
   10)15
      10
      ──
       5)10
         10
         ──
          0
```

Note that we do not have to record the quotients but only the remainders. An alternative way of writing this is as follows

$55 = 3 \times 15 + 10$
$15 = 1 \times 10 + 5$
$10 = 2 \times 5 + 0$ so gcd is 5

where the numbers that flow through are shown in various colours for clarity.

Although we don't need to record the quotients for finding the gcd, they will prove useful in the last lecture when we discuss cryptography and linear congruences.

The calculation of the gcd for the other two examples is shown opposite or can be written out thus

$252 = 2 \times 120 + 12$
$120 = 10 \times 12 + 0$ so gcd is 12

$44 = 2 \times 15 + 14$
$15 = 1 \times 14 + 1$
$14 = 14 \times 1 + 0$ so gcd is 1

The method can be illustrated geometrically by taking a rectangle whose sides are the numbers concerned and then considering its decomposition into squares as shown opposite.

```
120)252                                    15)44
   240         The gcd of 120 and             30
  12)120       252 is 12 (left).            14)15
     120                                       14
       0       The gcd of 15 and              1)14
               44 is 1 since they               14
               are relatively prime              0
                     (right).
```

In the first example below the large rectangle is 55 by 15. We take out three squares of side 15 and that leaves a rectangle 15 by 10. We take out one square of side 10 and that leaves a rectangle 10 by 5. And that can be subdivided into exactly two squares of side 5, so the gcd is 5. The same lurid colours are used as before when identifying the numbers that flow through.

The other examples show that the gcd of 44 and 15 is 1 and that the gcd of 120 and 252 is 12. It is important to appreciate that Euclid's algorithm is quite mechanical. We do not have to find the factors of the two numbers concerned.

Fibonacci numbers have some curious properties relating to the greatest common divisor. The first is simply that adjacent Fibonacci numbers are always relatively prime so that their gcd is 1. Thus gcd(5, 8) and gcd(55, 89) are equal

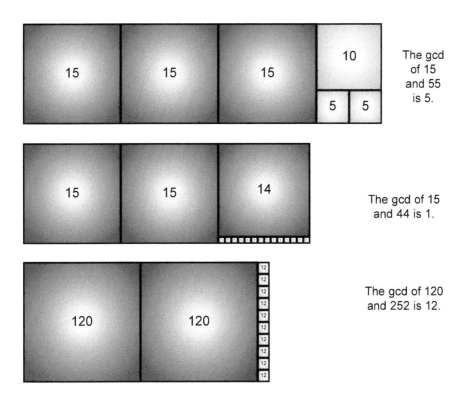

to 1. This is easily proved. Since $F_{n+1} - F_n = F_{n-1}$, it follows that if F_{n+1} and F_n have a common factor then it must also be a common factor of F_{n-1}. We can cascade this argument backwards and deduce that the factor must also be a factor of F_1. But F_1 is 1. So there can't be any common factors. That's it!

Even more surprising is that the gcd of any pair of Fibonacci numbers is itself always a Fibonacci number. For example

$$\gcd(8, 34) = 2 \quad \gcd(21, 144) = 3 \quad \gcd(55, 610) = 5$$

It can be shown that the gcd of F_m and F_n is F_g where g is $\gcd(m, n)$. So in the first example above F_6 is 8 and F_9 is 34. Now the gcd of 6 and 9 is 3 and F_3 is 2 which is consistent with the fact that the gcd of 8 and 34 is indeed 2.

A first step in proving this amazing fact about the gcd is to show that F_{mn} is always divisible by F_m and F_n. Thus in the case of $m = 3$ and $n = 5$ we find that $F_{15} = 610$ is divisible by both $F_3 = 2$ and $F_5 = 5$. The proof of this is not too hard and leads on to the amazing fact about the gcd.

The table below shows the gcd of all pairs of Fibonacci numbers less than 1000. Other facts are revealed by the table such as that every third number is divisible by $F_3 = 2$, every fourth is divisible by $F_4 = 3$ and so on.

Sometimes one Fibonacci number is a multiple of another in which case the gcd is simply the smaller. That happens with 377 which is a multiple of 13.

n	1	2	3	4	5	6	7	8	9	10	11	12	13	14	15	16
F_n	1	1	2	3	5	8	13	21	34	55	89	144	233	377	610	987
1	1	1	1	1	1	1	1	1	1	1	1	1	1	1	1	1
2		1	2	1	1	2	1	1	2	1	1	2	1	1	2	1
3			1	1	3	1	1	1	3	1	1	1	3	1	1	1	3
5				1	1	1	1	5	1	1	1	1	5	1	1	1	5	1
8					1	2	1	1	8	1	1	2	1	1	8	1	1	2	1
13						1	1	1	1	1	13	1	1	1	1	1	13	1	1
21							1	1	3	1	1	1	21	1	1	1	3	1	1	3
34								1	2	1	1	2	1	1	34	1	1	2	1	1	2	1
55									1	1	1	5	1	1	1	1	55	1	1	1	1	5	1
89										1	1	1	1	1	1	1	1	89	1	1	1	1	1
144											1	2	3	1	8	1	3	2	1	1	144	1	1	2	3
233												1	1	1	1	1	1	1	1	1	1	233	1	1	1
377													1	1	1	1	13	1	1	1	1	1	1	377	1	1
610														1	2	1	5	2	1	1	2	5	1	2	1	1	610	1
987															1	1	3	1	1	1	3	1	1	1	3	1	1	1	987

Table showing the gcd of pairs of Fibonacci numbers less than 1000.

Prime factors

THE PROBLEM of factorization, that is, finding the prime factors of a number is not straightforward. It is often known that a particular number is not prime, yet the factors are unknown. That happened with the Mersenne number M_{67} as mentioned in the lecture on Amicable Numbers.

The simplest technique is simply to try dividing the number by the known prime numbers. If the number is N then we only have to try primes less than $\sqrt{N}$ since if it has a prime greater than that it will also have a prime less than that.

So we need a list of primes. We can do this using the Sieve of Eratosthenes. Eratosthenes was a Greek mathematician in the 3rd century BC. For example to find all the primes less than 100, we first write out all the numbers up to 100. Ignore 1 and consider 2. Then cross out all those that are a multiple of 2. Then consider the next number not crossed out (it will be 3). Then cross out all multiples of 3 and so on. The first diagram below shows the stage just after all multiples of 3 have been crossed out.

For the purposes of illustration, those that are not crossed out are also in bold whereas those that have been crossed out are shown as normal. The second diagram shows the position when the exercise is completed. Actually we can stop after 7. The numbers left in bold are the prime numbers smaller than 100.

So in order to check a number less than 10000, we can just test for divisibility by all the primes less than 100. In the days of computers that is easy but with only a pencil it is tedious.

We know from the lecture on Notations that there are easy checks for divisibility by 2, 3, 5, and 11 and (by using 1001) also for 7 and 13. But for other primes it seems as if raw division of each one in turn is required. However, we can bundle them into groups and use Euclid's algorithm.

```
 1  2  3  4  5  6  7  8  9 10        1  2  3  4  5  6  7  8  9 10
11 12 13 14 15 16 17 18 19 20       11 12 13 14 15 16 17 18 19 20
21 22 23 24 25 26 27 28 29 30       21 22 23 24 25 26 27 28 29 30
31 32 33 34 35 36 37 38 39 40       31 32 33 34 35 36 37 38 39 40
41 42 43 44 45 46 47 48 49 50       41 42 43 44 45 46 47 48 49 50
51 52 53 54 55 56 57 58 59 60       51 52 53 54 55 56 57 58 59 60
61 62 63 64 65 66 67 68 69 70       61 62 63 64 65 66 67 68 69 70
71 72 73 74 75 76 77 78 79 80       71 72 73 74 75 76 77 78 79 80
81 82 83 84 85 86 87 88 89 90       81 82 83 84 85 86 87 88 89 90
91 92 93 94 95 96 97 98 99 100      91 92 93 94 95 96 97 98 99 100
```

Two stages in the sieve of Eratosthenes.

Suppose we want to check 15181. It is not divisible by 2, 3, 5, 7, 11, or 13 so it could be awkward. Let's check for divisibility by 17, 19, and 23 (the next three primes) by finding the gcd of 15181 and the product 17×19×23 which is 7429.

We get

```
7429)15181
     14858
     ‾‾‾‾‾
      323)7429
          7429
          ‾‾‾‾
             0
```

which can also be written as

$15181 = 2 \times 7429 + 323$
$7429 = 23 \times 323 + 0$ so gcd is 323

This means that 15181 is also divisible by $323 = 17 \times 19$. We then easily find that the other factor of 15181 is 47 so finally $15181 = 17 \times 19 \times 47$.

So one gcd calculation can find several primes at once. And we can bundle many primes together if we wish.

But this becomes particularly tedious if a number is simply the product of two primes such as $97 \times 103 = 9991$. Other techniques are necessary.

Fermat's method

FERMAT DISCOVERED a neat method of factorization. In essence, to find factors of n (assumed odd), we seek integral values x and y such that

$n = x^2 - y^2$ or equivalently $x^2 - n = y^2$

If we find such values then we can immediately factorize n since

$n = (x+y) \times (x-y)$

First we find the smallest k for which $k^2 > n$ and then we examine

$k^2-n, \ (k+1)^2-n, \ (k+2)^2-n$ and so on

until we find one that is a perfect square, say $(k+l)^2-n = y^2$. And then of course taking $x = k+l$, we have $x^2-n = y^2$ so that n can be factorized into $(x+y)\times(x-y)$.

It is important to realise that we will eventually find a value since we will try $x = (n+1)/2$ for which x^2-n is indeed a perfect square (it is the square of $(n-1)/2$) corresponding to the factorization $n = n \times 1$. If we get to this stage then we will have shown that n is prime. Note that n must be odd otherwise $(n+1)/2$ is not a whole number and the algorithm fails.

We will try this on 9991. The smallest k for which $k^2 > n$ is 100 ($100^2 = 10000$ and $99^2 = 9801$). So we try

$$100^2 - 9991 = 10000 - 9991 = 9 = 3^2$$

Gosh, we got a perfect square straight away! So $x = 100$ and $y = 3$ leading to

$$9991 = (100+3) \times (100-3) = 103 \times 97$$

That was all too easy. Let's try 8777 instead. The smallest k for which $k^2 > 8777$ is 94 whose square is 8836. So we try

$94^2 - 8777 = 8836 - 8777 = 59$	not a square
$95^2 - 8777 = 9025 - 8777 = 248$	not a square
$96^2 - 8777 = 9216 - 8777 = 439$	not a square
$97^2 - 8777 = 9409 - 8777 = 632$	not a square
$98^2 - 8777 = 9604 - 8777 = 827$	not a square
$99^2 - 8777 = 9801 - 8777 = 1024 = 32^2$	got a square!

So $x = 99$ and $y = 32$ leading to

$$8777 = (99+32) \times (99-32) = 131 \times 67$$

and we are done.

Computing the sequence of potential squares (59, 248, 439 etc.) looks tedious. However, the difference between pairs of adjacent squares increases by 2 each time so it is really quite easy. Thus

$59+189 = 248$, $\quad 248+191 = 439$, $\quad 439+193 = 632$, $\quad 632+195 = 827$, $827+197 = 1024$,

and we are done.

Checking whether a number is a square needs a bit of work sometimes. But we can note that the last digit of a perfect square cannot be 2, 3, 7, or 8 and that eliminates quite a few immediately. We can take that further and note that the last two digits of a perfect square can only be certain pairs of digits namely

00	01	04	09	16	21	24	25	29	36	41
44	49	56	61	64	69	76	81	84	89	96

and in this example that eliminates all the intermediate possibilities.

Now consider 8999. The smallest k such that $k^2 > 8999$ is 95. So we get

$95^2 - 8999 = 9025 - 8999 = 26$

$96^2 - 8999 = 9216 - 8999 = 217$

Neither of these are squares and their difference is 191, so we can now compute the others thus

217+193 = 410,   410+195 = 605,   605+197 = 802,   802+199 = 1001,
1001+201 = 1202,   1202+203 = 1405,   and so on

However, since 8999 is in fact prime we are going to have to try around 4400 numbers until we eventually get

$4500^2 - 8999 = 20250000 - 8999 = 20241001 = 4499^2$

from which 8999 = (4500+4499) × (4500−4499) = 8999 × 1 showing that 8999 is prime.

Clearly, Fermat's technique does not work so well when a number actually is prime. However, if we check for small primes first, say up to 37, then the search required is rather less. In the case of 8999, if it has no primes up to 37, then the largest prime factor it could have is of the order of 8999/41 or around 219. This would correspond to a possible factorization of 41 × 219 which corresponds to (130+89) × (130−89). So we would have to check squares from 95^2 to 130^2 which is not too bad. Just 36 numbers rather than 4400.

Eratosthenes revisited

THE SIEVE of Eratosthenes can be considered in a different way. Imagine a number of people each representing a prime number and whose job is to eliminate all numbers that they divide. They are linked together in order.

We start with a person whose task is to throw away all even numbers (the 2-person). Some overall person in charge (the Boss) gives to the 2-person cards inscribed with the integers in order. The 2-person discards all even numbers and passes the odd cards to the adjacent 3-person. The 3-person discards all cards divisible by 3 and passes the others to the adjacent 5-person. And so on.

The system grows because whenever an N-person has a number M to pass on and finds that there is no next person, then a next person is allocated and becomes the M-person. The numbers allocated to the persons are of course the prime numbers.

In the diagram opposite, the Boss is about to give 10 to Mr 2. The next number, 11, awaits on the ground. Mr 2 has previously thrown away 4, 6, and 8. Mr 2 has just given 9 to Mr 3 and Mr 3 is about to throw it away. Mr 5 is holding 7 recently given to him by Mr 3 and is deciding what to do about it. He has

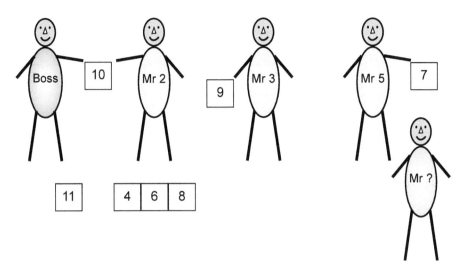

The human sieve of Eratosthenes.

decided that he does not divide into it and so wishes to give it to a new person who will become Mr 7. A new person awaits below.

This is easy to simulate using a programming language with inherent multiple processes.

One reason for introducing this alternative view of the sieve of Eratosthenes is that we want to consider other systems of numbers other than simple integers which support the concept of primes.

Complex numbers

IN THE LECTURE on Fractions, we mentioned that the positive real numbers could be categorized as

Rational numbers which can be integers or fractions, and

Irrational numbers which can be algebraic or transcendental.

In the lecture on Notations, we saw the importance of zero in place notation.

But the positive numbers and zero are not enough because we naturally ask what is the meaning of 5–7 or perhaps what is the solution of equations such as

$$x + 1 = 0$$

Moreover, it seems wrong that some quadratic equations such as

$$x^2 - 7x + 12 = 0$$

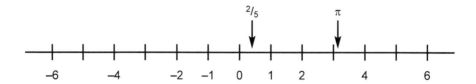

The real line showing various numbers.

have two roots (3 and 4) but others such as

$$x^2 + x - 12 = 0$$

seem to have only one (positive) root (3).

We know that the answer lies in the existence of negative numbers as well as positive numbers. In fact we can think of the real numbers as being spread along an infinite line with zero in the middle as shown above.

We then have to introduce rules for dealing with negative numbers such as that multiplying two negative numbers together gives a positive one. Thus we obtain $2 \times 2 = 4$ and also $-2 \times -2 = 4$. So 4 has two square roots, +2 and −2.

But we are still not satisfied since we now have problems such as what is the square root of −4 and why do some quadratic equations such as

$$x^2 + 2x + 2 = 0$$

appear to have no roots at all?

The line of real numbers is not enough and we introduce a whole new dimension of numbers around the concept of a special number usually denoted by i which has the important property that $i^2 = -1$. All the normal rules apply to i. Thus we can multiply i by 2 to get $2i$ and so on. Such numbers are often called imaginary numbers. Moreover, we can add 3 to $2i$ to give a number which we simply denote by $3+2i$ which is an example of a general complex number.

These numbers and the real numbers can be depicted as lying in a plane − the complex plane or Argand plane after the Swiss mathematician Jean Robert Argand (1768−1822). The real numbers continue to lie on a horizontal line, the x-axis or real axis and the pure imaginary numbers lie on a vertical line, the y-axis or imaginary axis.

Working with complex numbers is very easy. The operations are just as with real numbers except that whenever we get i^2, we replace it by −1.

For example, consider addition

$$(6+3i) + (2-i) = (8+2i)$$

We just add the real parts together giving $6 + 2 = 8$ and similarly add the imaginary parts giving $3i - i = 2i$.

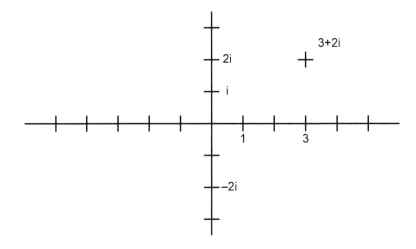

The complex or Argand plane.

Multiplication is a bit more interesting, for example

$$(6+2i) \times (2-i) = 12 + 4i - 6i - 2i^2 = 12 - 2i + 2 = (14-2i)$$

where we had to replace $2i^2$ by -2.

An important idea is that of a complex conjugate. The conjugate of $(x+iy)$ is simply $(x-iy)$ – we just change the sign of i.

Note carefully that any number times its conjugate is always real. Thus

$$(3+4i) \times (3-4i) = 9 + 12i - 12i - 16i^2 = 9 + 16 = 25.$$

This is a useful trick for division. To divide one complex number by another, we multiply both numbers by the conjugate of the second and then do the division which is then very easy. So to divide $(4+2i)$ by $(1-2i)$, we first do

$$(4+2i) \times (1+2i) = 4 + 8i + 2i - 4 = 10i$$

$$(1-2i) \times (1+2i) = 5$$

and so the answer is

$$(4+2i) / (1-2i) = 10i / 5 = 2i$$

which can easily be checked by multiplying out again.

Having discovered that the real numbers were not complete in some sense, we naturally ask whether the complex numbers are complete. For example, can we find the square root of every complex number and are the roots of every

172 Nice Numbers

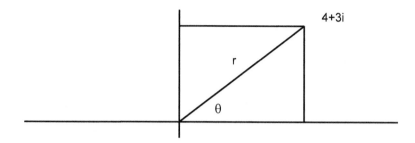

Cartesian and polar notation.

equation with complex coefficients expressible as complex numbers? The answer is yes, the complex number system is algebraically complete.

We have been looking at complex numbers essentially as x- and y-cartesian coordinates in the complex plane. (Note cartesian after the French mathematician René Descartes (1596–1650).) An alternative view is to consider polar coordinates as shown above. Thus a value such as $4+3i$ can alternatively be described in terms of the distance r and the angle θ. In the case of $4+3i$, the value of r is of course 5 since by Pythagoras we have a right angled triangle and $5^2 = 3^2+4^2$ and the angle θ is given by $\sin \theta = 3/5$.

An important property of this view is that if we multiply two numbers then the length r of the result is the product of the lengths of the two numbers and the angle of the result is the sum of the angles of the two numbers.

So to find the square roots of a complex number we take the square root of the length and halve the angle. Clearly this can be done with every complex number. As an example, the square roots of $2i$ are $(1+i)$ and $(-1-i)$. Remember that every number has two square roots.

Other operations such as taking powers and logarithms can also be applied to complex numbers. For example, recall the series for e (the base of natural logarithms which we met in the lecture on Probability), and the trigonometric functions, sine and cosine, which are

$$e^x = 1 + \frac{x}{1!} + \frac{x^2}{2!} + \frac{x^3}{3!} + \dots$$

$$\sin x = \frac{x}{1!} - \frac{x^3}{3!} + \frac{x^5}{5!} - \dots$$

$$\cos x = 1 - \frac{x^2}{2!} + \frac{x^4}{4!} - \dots$$

Now, if we put $x = i\theta$ in the series for e^x, we get

$$e^{i\theta} = 1 + \frac{i\theta}{1!} - \frac{\theta^2}{2!} - \frac{i\theta^3}{3!} + \frac{\theta^4}{4!} - \dots$$

$$= \cos \theta + i \sin \theta$$

This leads to some very curious results. First of all put $\theta = \pi$ and remembering that $\sin \pi = 0$ and $\cos \pi = -1$, this gives

$$e^{\pi i} = \cos \pi + i \sin \pi = -1$$

which is a very strange relationship between the three numbers e, π, and i.

If we put $\theta = \pi/2$ and in this case noting that $\sin \pi/2 = 1$ and $\cos \pi/2 = 0$, we get

$$e^{\pi i/2} = \cos \pi/2 + i \sin \pi/2 = i$$

Now raise both sides to the power i

$$i^i = (e^{\pi i/2})^i = e^{-\pi/2} = 0.2078795...$$

So, surprisingly, the value of the imaginary number i raised to the power i is actually a real number. That is surely a candidate for being somebody's favourite number!

Complex primes

COMPLEX NUMBERS of the form $a+ib$ where a and b are integers are often called Gaussian integers.

We define a complex prime by analogy with ordinary primes. A complex integer is prime if its only factors (in terms of complex integers which includes real integers) are itself and 1. It turns out that factors of i (and -1) have to be ignored if factorization is to be unique.

For example $(3+i)$ is not a prime since it is the product of $(1-i)$ and $(1+2i)$. Note moreover that $(1-i) \times i$ is $(1+i)$. This means that we can work just with primes whose real and imaginary parts are both positive. So finally we factorize $(3+i)$ thus

$$(3+i) = -i(1+i)(1+2i)$$

Some real integers which are normally prime such as 2 and 5 are not prime as complex numbers. Thus

$$2 = (1+i)(1-i) = -i(1+i)^2$$
$$5 = (1+2i)(1-2i) = -i(1+2i)(2+i)$$

In a similar way, 13 and 17 are not prime complex numbers either, since

$$13 = -i(2+3i)(3+2i)$$
$$17 = (1+4i)(1-4i) = -i(1+4i)(4+i)$$

174   Nice Numbers

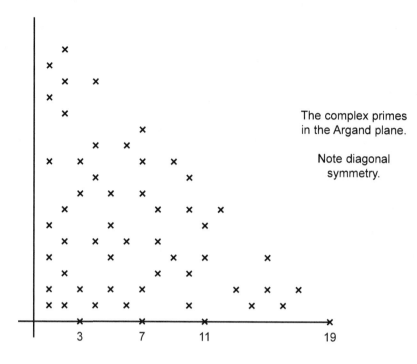

The complex primes in the Argand plane.

Note diagonal symmetry.

However, 3, 7, 11, and 19 are indeed prime complex numbers.

Here is a list of the first few complex primes. Each row contains those primes whose real and imaginary parts add to the same

$1+1i$							
$1+2i$	$2+1i$	3					
$1+4i$	$2+3i$	$3+2i$	$4+1i$				
$1+6i$	$2+5i$	$5+2i$	$6+1i$	7			
$2+7i$	$4+5i$	$5+4i$	$7+2i$				
$1+10i$	$3+8i$	$5+6i$	$6+5i$	$8+3i$	$10+1i$	11	
$3+10i$	$4+9i$	$5+8i$	$8+5i$	$9+4i$	$10+3i$		
$1+14i$	$2+13i$	$4+11i$	$7+8i$	$8+7i$	$11+4i$	$13+2i$	$14+1i$
$1+16i$	$2+15i$	$6+11i$	$7+10i$	$10+7i$	$11+6i$	$15+2i$	$16+1i$
$2+17i$	$4+15i$	$7+12i$	$9+10i$	$10+9i$	$12+7i$	$15+4i$	$17+2i$	19

Note that if $a+ib$ is a prime then so is $b+ia$. In other words they are reflected in the diagonal of the positive quadrant of the Argand plane as shown in the diagram above.

In order to compute the complex primes we can use the principle of the Sieve of Eratosthenes.

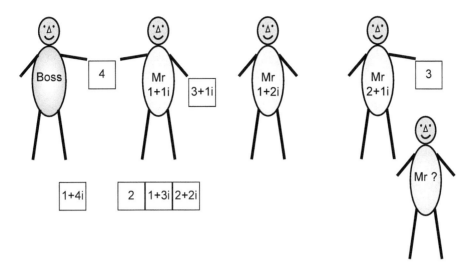

The human sieve of Eratosthenes for complex primes.

The Boss now hands out the complex integers in a reasonable order and the rest of the algorithm is much as before. An important issue is what is a reasonable order? The key point is that small ones must come first, but we don't have a unique ordering for complex integers. But we can order them roughly by modulus (the length r in polar form). We don't have to accurately follow that order provided we are assured that a number does not occur before any of its factors. Luckily the order obtained by doing them in groups where the sum of the imaginary and real parts are the same and then doing the groups in ascending order of the sum does just this. So we present the numbers in the order

$1+1i$	2				
$1+2i$	$2+1i$	3			
$1+3i$	$2+2i$	$3+1i$	4		
$1+4i$	$2+3i$	$3+2i$	$4+1i$	5	
$1+5i$	$2+4i$	$3+3i$	$4+2i$	$5+1i$	6

and so on.

In the diagram above, three primes have been found and the Boss is about to give 4 to Mr $1+1i$. The next number, $1+4i$, awaits on the ground. Mr $1+1i$ has previously thrown away 2, $1+3i$, and $2+2i$ and is about to throw away $3+1i$. Mr $1+2i$ is idle at the moment. Mr $2+1i$ is holding 3 recently given to him by Mr $1+2i$ and is deciding what to do about it. He has decided that he does not divide into it and so wishes to give it to a new person who will become Mr 3. A new person awaits below.

Note that Mr 1+2*i* has not yet been given anything that he can throw away. Everything has so far been thrown away by Mr 1+1*i*. Moreover, Mr 1+1*i* is going to throw away the next number, 4. The next few numbers will all be found to be prime. But Mr 1+2*i* will throw away 5.

Polynomials

WE ARE FAMILIAR with the general idea of a polynomial consisting of the sum of a number of terms of different powers of some variable x such as

$x^2 + 5x + 6$

If we restrict the coefficients to be real integers such as 5 and 6 then we can consider which polynomials can be written as products of other polynomials and which cannot. Those that cannot are prime polynomials.

As an example, the above polynomial can be factorized thus

$x^2 + 5x + 6 = (x + 2) \times (x + 3)$

and so is not a prime polynomial. On the other hand

$x^2 + x + 1$

cannot be factored and so is prime.

Note that in this model, all polynomials of the form

$x + n \qquad n = 1, 2, 3, ...$

are prime. And of course there are an infinite number of such linear primes which is perhaps a bit boring.

We can reduce the number of polynomials to be considered by supposing that we work using modular arithmetic for the coefficients using some base. Thus if we worked in base 4 then the only linear polynomials are the four

$x+0, \quad x+1, \quad x+2, \quad x+3$

and there are 16 quadratic polynomials namely

$x^2, \ x^2+1, \ x^2+2, \ x^2+3, \ x^2+x, \ x^2+x+1, \ x^2+x+2, \ x^2+x+3, \ x^2+2x,$
$x^2+2x+1, \ x^2+2x+2, \ x^2+2x+3, \ x^2+3x, \ x^2+3x+1, \ x^2+3x+2, \ x^2+3x+3$

Note that if we multiply $x+2$ by $x+3$, we get x^2+x+2. This is because we would get x^2+5x+6 in normal arithmetic. But 5 reduces to 1 mod 4 and 6 reduces to 2.

So x^2+x+2 which is prime using normal integer coefficients is not prime in the mod 4 world.

A particularly strange set of polynomials is where all the coefficients are considered to be modular with base 2. So all coefficients are either zero or one. The first few polynomials are then

$x, \quad x+1$

$x^2, \quad x^2+1, \quad x^2+x, \quad x^2+x+1$

$x^3, \quad x^3+1, \quad x^3+x, \quad x^3+x+1, \quad x^3+x^2, \quad x^3+x^2+1, \quad x^3+x^2+x, \quad x^3+x^2+x+1$

And the first few prime ones are

$x, \quad x+1$

x^2+x+1

$x^3+x+1, \quad x^3+x^2+1$

$x^4+x+1, \quad x^4+x^3+1, \quad x^4+x^3+x^2+x+1$

$x^5+x^2+1, \quad x^5+x^3+1, \quad x^5+x^3+x^2+x+1, \quad x^5+x^4+x^2+x+1, \quad x^5+x^4+x^3+x+1,$
$x^5+x^4+x^3+x^2+1$

$x^6+x+1, \quad x^6+x^3+1, \quad x^6+x^4+x^2+x+1, \quad x^6+x^4+x^3+x+1, \quad x^6+x^5+1$

The reader will appreciate that this topic can be extended in various ways.

Further reading

PRIME FACTORIZATION, the greatest common divisor, and Fibonacci numbers are admirably discussed in *Elementary Number Theory* by David Burton. A more advanced text is *Prime Numbers and Computer Methods for Factorization* by Hans Riesel. The dynamic method for the Sieve of Eratosthenes and its use with Gaussian primes and polynomials are described in *Programming in Ada 2012* by the author.

A good read is *An Imaginary Tale – the story of $\sqrt{-1}$* by Paul Nahin.

Exercises

1 Find the gcd of 7429 and 13889. Hence find the prime factors of 13889.

2 Using Fermat's method find the prime factors of 12319.

3 Solve the quadratic equation
$x^2 + 2ix - (1+2i) = 0$

9 Music

THIS LECTURE looks at the evolution of musical scales from the Pythagoreans to modern times. It explores the conflicts inherent in attempting to get harmonious scales in many keys using a limited keyboard. The story starts with the Pythagorean scale which seems a very natural subdivision of the octave but is difficult for polyphonic choirs and led to just intonation. But this is difficult if scales in many keys are required and was followed by forms of meantone which was especially used on organs. Finally, we move to the compromise of the equitempered scale almost universally used today.

Frequency and vibrations

SOUND IS CAUSED by vibrations and their frequency determines the pitch of a sound. When a taut string vibrates it has a natural frequency which depends upon its tension, length, and mass per unit length. If the tension is doubled the frequency is doubled and if the length is doubled then the frequency is halved. So we have

$f = kT/l$ frequency in terms of tension T and length l

where k is a constant depending on factors such as the mass.

The principal mode of vibration of a string is where the string moves as a simple loop with the centre of the string moving most. Suppose the frequency of this mode is f. Another possible mode of vibration is where the string vibrates as two sections with the centre actually still. If the tension is unchanged then the frequency will be doubled because the length of each part is halved – so this mode has frequency $2f$. The string can also vibrate in three parts with frequency $3f$ and so on. These various modes of vibration are depicted below.

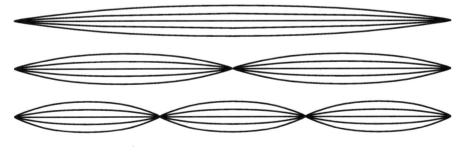

Modes of vibration of a string.

So altogether a string can vibrate with frequencies f, $2f$, $3f$, $4f$, ... The higher frequencies are called harmonics or overtones of the principal frequency f. Similar behaviour occurs with the vibration of air in a tube. Again there is a principal frequency f and various harmonics $2f$, $3f$, and so on.

Most instruments in the orchestra make their sounds by vibrating strings (the strings obviously such as violin and cello, plus piano and harp) or by vibrating air in pipes (the woodwind such as oboe and clarinet and the brass such as horn and trombone, plus the organ).

The reason why different instruments sound different is largely because the proportions of the various harmonics are different. There is another reason as well. In some instruments a note is continuous such as in the case of an organ whereas in others the sound decays away as in a piano. Even in the case of an organ, there will be initial attack transients which fade leaving the steady note remaining.

However, many of the percussion instruments such as the drums and gong vibrate differently. A circular drum has a principal mode whereby the skin moves in a uniform manner with the centre moving most and all other points moving according to their distance from the centre. But there are many other possible modes of vibration and their frequencies do not have a simple relationship with the principal frequency. The result is a bit of a mess of sounds and that is why such instruments do not appear to have a well-defined musical note.

The simple ratios of 1:2 and 2:3 derived from f, $2f$, $3f$ form the basis for developing harmony and scales. Thus if two notes have frequency f and $2f$ then they are an octave apart. For example if middle C on a piano were 240 Hz (cycles per second), then the C above (C') would be 480 Hz (these are not the correct values but are convenient for illustration because 240 has helpful factors – the correct values are about 10% higher). If two notes have a frequency ratio of 3:2 then they are said to form a perfect fifth. So again if middle C has frequency 240, then the G above will have frequency of 240×3/2 = 360.

The terms fifth and octave derive from the fact that in counting the notes in an interval both ends are counted. CDEFG has five notes and so the interval CG is a fifth. And an octave has eight notes CDEFGABC'.

The interval from G to C' is a perfect fourth and this has ratio of 4:3. If we take a fifth down from C' then we get to F – if C' is 480 then F is 480×2/3 = 320. From C to F is also a ratio of 4:3 and is another perfect fourth.

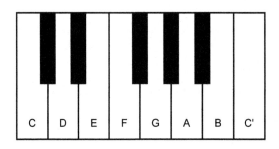

CC' is an octave.

CG and FC' are fifths.

CF and GC' are fourths.

The interval from C to E is known as a third and that from C to A is a sixth, similarly that from C to D is a second and so on. The intervals of a fourth and a fifth are known as perfect since they are the most harmonious. This is because of the simplicity of the ratios 3:2 and 4:3. We will now look at different ways of completing the scale and assigning frequencies to the remaining notes D, E, A, and B.

The Pythagorean scale

THE RATIO between F and G is 360/320 or 9/8. This is an interval known as a tone (strictly a major tone). If we take a fourth down from G, we obtain D which will be a tone up from C and so have frequency 240×9/8 = 270. Taking a fifth up from D gives A at 270×3/2 = 405.

There are two ways of getting E and B. We can simply say that DE and AB should also be tones of ratio 9/8. This gives E at 270×9/8 = 303.75 and B at 405×9/8 = 455.625. B of course is simply a fifth above E. Alternatively we can simply keep going up by fifths from A and then transpose down an octave as necessary. A fifth above A is E' at 405×3/2 = 607.5 and then we halve this to give E = 303.75 as before. And then, as we have just noted, B is a fifth above E.

We have now completed the Pythagorean scale of C and can depict it as below. Note that some new intervals have been introduced such as the major third CE of ratio 81/64, the major sixth CA of ratio 27/16 and the major seventh of ratio 243/128.

The small intervals EF and BC are both equal to a fourth diminished by two tones and so are 4/3×8/9×8/9 = 256/243 = 1.053 .... This small interval is known as a semitone – but note carefully that it is slightly smaller than half a tone of 9/8 which would be √(9/8) = 1.060 .... The semitone of 256/243 is also known as the

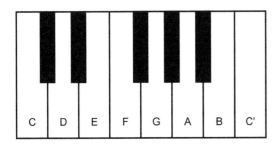

The Pythagorean scale of C showing frequencies and ratios.

C	D	E	F	G	A	B	C'	
240	270	303³/₄	320	360	405	455⁵/₈	480	the frequencies
1/1	9/8	81/64	4/3	3/2	27/16	²⁴³/₁₂₈	2/1	ratios from C
	9/8	9/8	²⁵⁶/₂₄₃	9/8	9/8	9/8	²⁵⁶/₂₄₃	ratios between notes

Pythagorean Limma. The difference between two of these semitones and the major tone is a ratio of 531441/524288 and is known as the Pythagorean Comma. The reader should note that we are already encountering weird ratios; there is more trouble ahead!

We now turn to transposing to other scales which introduces the black notes for sharps and flats.

Suppose we start the scale at G (a fifth above C). We find that the sequence GABCDE has exactly the same ratios between adjacent notes as the sequence CDEFGA. But EF does not match AB since EF is a semitone and AB is a tone. So we introduce one new note which we call F sharp (and write F#) whose frequency is 9/8 times that of E (and so $341^{23}/_{32}$). We thus get the scale of G whose notes are GABCDEF#G. Note that the intervals FF# and F#G are *not* the same. FF# is 2187/2048 and is known as the Pythagorean Apotome or chromatic semitone whereas F#G is 256/243, the Pythagorean Limma or diatonic semitone.

There seems to be confusion regarding the names of the semitones. Some claim that the chromatic semitone is so called because it is the difference between a note and its sharp or flat (that is a coloured note) as in the case of FF# whereas the diatonic semitone is that between two notes of different names such as EF. However, beware that the numerical ratios will depend on the way in which the scale is derived.

Similarly, if we start at D (a fifth above G) then again everything matches the scale starting at G except that we have to introduce a second sharp namely C sharp whose frequency is 9/8 times that of B. So we get the scale DEF#GABC#D. And then starting at A (a fifth above D) we have to introduce a third sharp namely G sharp whose frequency is 9/8 times that of F sharp and that gives the scale ABC#DEF#G#A. We can continue in this way introducing sharps and they all divide the tone concerned into two unequal semitones, the first the chromatic semitone of 2187/2048 and then the diatonic semitone of 256/243.

We can now turn to the flats. Consider the scale starting at F (a fifth below C). It matches the scale starting at C except that we have to introduce B flat (which we write as B♭) whose frequency is 8/9 times that of C' (and so $426^{2}/_{3}$). We get the scale FGAB♭CDEF. Note again that the intervals AB♭ and B♭B are not the same. AB♭ is the diatonic semitone of 256/243 and B♭B is the chromatic semitone of 2187/2048.

And then we consider starting a fifth below F which is B♭. This introduces E flat whose frequency is 8/9 times that of F and we get the scale B♭CDE♭FGAB♭. Now start a fifth below B♭ which is E♭. This introduces A flat whose frequency is 8/9 times that of B flat and gives the scale E♭FGA♭B♭CDE♭. And so on.

So the flats again divide the tones into two semitones as the sharps did but this time the division is in reverse in each case. On a traditional keyboard, a single key is provided that has to do duty as both C# and D♭ for example. So really the black keys need to be duplicated in order to provide the correct notes for the Pythagorean scale. The difference between G# and A♭ is the difference between a major tone of 9/8 and two diatonic semitones of 256/243, and as

The pattern of flats and sharps.

mentioned earlier this is 531441/524288, the Pythagorean Comma. Note that the flats are lower than the corresponding sharps. The pattern is shown above.

Another way of computing the difference is to note that we got to G# by a sequence of four whole tones up from C. So its frequency is $(9/8)^4$ times that of C. And we got to A♭ by a sequence of two whole tones down from C'. So its frequency is $(8/9)^2$ times that of C'. Clearly these are not the same. Reducing everything to powers of 2 and 3 we have

G# is $3^8/2^{12}$ times that of C and A♭ is $2^7/3^4$ times that of C

and these are not the same since 2^{19} (= 524288) is not the same as 3^{12} (= 531441). The Pythagorean Comma is quite small, the ratio $3^{12}/2^{19}$ being 1.01364....

If we have only one set of black keys then the obvious thing to do is to tune them to give the correct values for F#, C#, B♭, and E♭ and to compromise on G#/A♭. This means that scales of one and two sharps and flats are perfect, three sharps and flats are not quite right but more extreme keys are decidedly wrong.

The overall conclusion is that the Pythagorean scale works quite well provided we do not stray too far from the scale of C.

In early music, intervals mostly occurred as sequences rather than as chords, and in those cases where two notes were played or sung together they were inevitably a fifth or an octave apart. Fifths and octaves have simple ratios and the human ear likes simple ratios. However, the ratio 81/64 for the major third is not simple and multipart music involving thirds (and sixths such as CA) sounds rough using the Pythagorean scale. So an improvement was sought and this led to just intonation which we will explore in a moment.

Cents

SINCE MUSICAL INTERVALS are described in terms of ratios it is convenient to use a logarithmic scale for their description and comparison. A cent is defined such that an octave is 1200 cents. Since an octave has 12 semitones, it follows that if the semitones are all equal then they will be exactly 100 cents and a tone comprising two semitones will be 200 cents and so on. The intervals we have met so far are given in the table overleaf.

cents	ratio		name
24	531441:524288	$3^{12}:2^{19}$	Pythagorean Comma
90	256:243	$2^8:3^5$	Pythagorean Limma, diatonic semitone
114	2187:2048	$3^7:2^{11}$	Pythagorean Apotome, chromatic semi
204	9:8	$3^2:2^3$	Pythagorean or major tone
408	81:64	$3^4:2^6$	Pythagorean major third
498	4:3	$2^2:3^1$	perfect fourth
702	3:2	$3^1:2^1$	perfect fifth
906	27:16	$3^3:2^4$	Pythagorean major sixth
1200	2:1		octave

Intervals in the Pythagorean system in cents.

Note how the intervals are all simply the ratios of a power of 2 to a power of 3. Cents are useful for judging whether an interval is acceptable or not. Thus the Pythagorean major third is 408 cents whereas, as we shall see in a moment, a just major third of ratio 5/4 is 386 cents. This difference of 22 cents is more than can be accepted for multipart music and quantifies the problem with the Pythagorean scale. As a general rule of thumb, an error of more than 15 cents is not acceptable.

Here is the complete Pythagorean scale covering from 5 flats to 5 sharps giving the frequencies and cents for all notes

C	240	0
D♭	$252^{68}/_{81}$	90
C#	$256^{37}/_{128}$	114
D	270	204
E♭	$284^4/_9$	294
D#	$288^{333}/_{1024}$	318
E	$303^3/_4$	408
F	320	498
G♭	$337^{29}/_{243}$	588
F#	$341^{23}/_{32}$	612
G	360	702
A♭	$379^7/_{27}$	792
G#	$384^{111}/_{256}$	816
A	405	906
B♭	$426^2/_3$	996
A#	$432^{999}/_{2048}$	1020
B	$455^5/_8$	1110
C	480	1200

We now turn to just intonation with its aim of perfect harmony.

Just intonation

JUST INTONATION makes the major third CE have a ratio of 5/4. As a consequence the notes of the familiar major chord CEGC' have frequency ratios 4:5:6:8 and these small numbers explain why this chord sounds so harmonious in just intonation.

Moreover, making the major third 5/4 and keeping the major tone CD at 9/8 means that DE becomes a minor tone of ratio 10/9. And the semitone EF which was the peculiar 256/243 in the Pythagorean scale becomes the much simpler 16/15 in just intonation. It is accordingly often called the just semitone.

The ratio of the major tone to the minor tone becomes 9/8 to 10/9 which is 81/80 and is known as the Syntonic Comma. This is also the ratio between the Pythagorean major third of 81/64 and the just major third of 5/4 as was mentioned earlier.

Beware that the Syntonic Comma is 22 cents and, as we have already noted, an error of more than 15 cents is not generally acceptable and so we must be vary wary of the Syntonic Comma!

It is interesting to note that the recurring decimal fraction for 80/81 is the curious expansion

   0.987654320987654320...

as was mentioned in the lecture on Fractions.

The resulting new frequencies and ratios for the scale of C then become as shown below. They are so much simpler that clearly this new scale must be a great improvement. The frequencies are all whole numbers and all the ratios are simple.

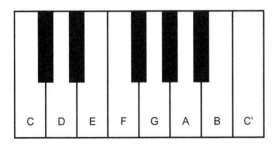

The just scale of C showing frequencies and ratios.

f C$_{''}$	60	$7f$?	420
$2f$ C$_{'}$	120	$8f$ C'	480
$3f$ G$_{'}$	180	$9f$ D'	540
$4f$ C	240	$10f$ E'	600
$5f$ E	300	$11f$?	660
$6f$ G	360	$12f$ G'	720

The harmonics of the low C of 60.

Another way of looking at this new scale is to consider the harmonics of a low C of frequency 60. They are shown in the table above. Apart from the 7th and 11th harmonics, all the harmonics are exactly present in the just scale. Note especially that D fits in because CD is a major tone and DE is a minor tone and not vice versa.

The scale of C major can be summarized thus

C	240	do	tonic
D	270	ra	supertonic
E	300	mi	mediant
F	320	fa	subdominant
G	360	so	dominant
A	400	la	submediant
B	450	ti	leading note
C'	480	do	tonic

which also gives the traditional names for the notes.

It is interesting to compare the Pythagorean and just scales and to note the difference thus

note	Pythagorean	just	ratio	cents
C	240	240	1:1	0
D	270	270	1:1	0
E	$303^3/_4$	300	81:80	22
F	320	320	1:1	0
G	360	360	1:1	0
A	405	400	81:80	22
B	$455^5/_8$	450	81:80	22
C	480	480	1:1	0

So E, A, and B all differ by a syntonic comma.

It is now time to consider transposition to other scales. Sadly, we get into serious trouble very quickly.

As before, consider the scale starting at G. Not only do we need F# but we find that A is not correct. The interval CD is a major tone but the GA of the just scale in C is a minor tone. We really need A to be 405 (ironically the same as the Pythagorean A). The two versions of A differ by a syntonic comma of 22 cents. We can refer to this higher A as A+. Indeed every transposition introduces two new notes – a black note as expected but also a new version of an existing white note differing by a syntonic comma from the normal one.

There were two approaches to solving this. One was simply to put up with it and to avoid keys far from C. But remember that it was the error of a syntonic comma in the major third that drove us from the Pythagorean scale in the first place. If one strayed far from C in the just scale, then more intervals were wrong and the result was awful. Let's explore that in more detail by considering the transpositions up and down and calculate frequencies for the black notes.

F sharp needs to be a major tone (9/8) above E (300) and equivalently a semitone (16/15) below G (360). This give $337\frac{1}{2}$. And then C sharp needs to be a semitone below D (270) or a major tone above B (450/2). This is $253\frac{1}{8}$. And G sharp needs to be a semitone below A (400) or a major tone above F sharp ($337\frac{1}{2}$). The latter is $379\frac{11}{16}$ whereas the former is 375. The discrepancy arises because we should have used A+ of 405 for the sharp scales. Clearly G# should therefore be $379\frac{11}{16}$. Incidentally, we also really need E+ and B+.

And now let's consider the flats. B flat needs to be a semitone above A (400) or a major tone below C' (480). This is $426\frac{2}{3}$. And then E flat needs to be a semitone above D (270) or a major tone below F (320). The former is 288 but the latter is $284\frac{4}{9}$. That is because D is wrong for the flat scales just as A was wrong for the sharp scales. D should be D– of $266\frac{2}{3}$, the difference again being a syntonic comma but in the opposite direction. So E flat should be $284\frac{4}{9}$. And finally A flat needs to be a semitone above G (360) or a major tone below B flat ($426\frac{2}{3}$). The former is 384 and the latter is $379\frac{7}{27}$. Again we get a discrepancy because we should be using G– for this scale.

So it is all a bit of a mess. We can just about agree on reasonable values for F#, C#, B♭, and E♭. And for G# alias A♭ we can compromise between $379\frac{11}{16}$ and $379\frac{7}{27}$ which are remarkably close and perhaps settle for say $379\frac{1}{2}$. So we can tune a traditional piano as shown in the table below.

If we now consider the fifths in the various keys we find that they are all very good except for D to A which is in error by a syntonic comma of 22 cents. And the fourths are similarly good except for A to D. But many of the major

C	240	G	360
C#	$253\frac{1}{8}$	G#/A♭	$379\frac{1}{2}$
D	270	A	400
E♭	$284\frac{4}{9}$	B♭	$426\frac{2}{3}$
E	300	B	450
F	320	C'	480
F#	$337\frac{1}{2}$		

Possible compromise tuning in just intonation.

thirds are not good. In fact only CE, DF#, FA, and GB are correct. All the others are bad. This is ironic because a key reason for introducing just intonation was to improve the major thirds. As a consequence scales far from C are bad.

If we move to G with one sharp then the thirds become GB, CE, and DF# and they are correct. But if we move to D with two sharps we find DF#, GB, and AC# of which AC# is incorrect (by a syntonic comma). And if we move to A with three sharps, we have AC#, DF#, and EG# and two of these are incorrect. The problems are worse with the flats. In the case of F with one flat the thirds are FA, B♭D, and CE and even B♭D is incorrect. So one was really forced to use only those keys very close to C.

The full range of ideal frequencies for all keys from 6 flats to 6 sharps is shown in the table opposite. As we go up the table, the value of a note changes by the factor 81/80 (the syntonic comma) from time to time. It is interesting to note that the difference between pairs such as C# and D♭ is very small. The ratio is in fact 32805:32768 and is known as the Skhisma and is about 2 cents.

We have already mentioned that one approach to the problem was just to ignore it and avoid keys far from C. The other approach to the problem was to add other physical keys to the keyboard. If we wish for precise just intonation in all keys from 6 flats to 6 sharps, then every white key has to be duplicated and every black key has to be triplicated and we also have to add keys for E# and C♭.

Altogether 31 keys are needed per octave as opposed to the normal 12 with frequencies as shown in the table below. Such keyboards were built and it is reported that Handel played a 31-note organ in the Netherlands (it is not clear whether the 31 notes on Handel's organ were the 31 notes in the table).

A keyboard with 31 keys was described by Marin Mersenne whom we have met in conjunction with Mersenne primes. This is discussed in Appendix J but as we shall see they are not the 31 notes we have just identified here.

However, if we are prepared to ignore the Skhisma and use only keys from 3 sharps to 4 flats then single black keys suffice and it is only the white keys that need to be duplicated a syntonic comma apart. To do this requires 19 keys per octave. Again a 19-note keyboard is mentioned in Appendix J.

C	240	$237^1/_{27}$	G♭	$337^{29}/_{243}$		
C#	$253^1/_8$	$256^{37}/_{128}$	G	360	$355^5/_9$	
D♭	$252^{68}/_{81}$		G#	$379^{11}/_{16}$	$384^{111}/_{256}$	
D	270	$266^2/_3$	A♭	$379^7/_{27}$		
D#	$284^{49}/_{64}$		A	400	405	
E♭	$284^4/_9$	$280^{680}/_{729}$	A#	$427^{19}/_{128}$		
E	300	$303^3/_4$	B♭	$426^2/_3$	$421^{97}/_{243}$	
E#	$320^{185}/_{512}$		B	450	$455^5/_8$	
F	320	$316^4/_{81}$	C♭	$449^{359}/_{729}$		
F#	$337^1/_2$	$341^{23}/_{32}$	C'	480		

The 31 notes need for just intonation from 6 flats to 6 sharps.

Key	C	C#	D	D#	E	E#	F#	G	G#	A	A#	B	C'
F#, 6#		$256\frac{37}{128}$		$284\frac{49}{64}$		$320\frac{185}{512}$	$341\frac{23}{32}$		$384\frac{111}{256}$		$427\frac{19}{128}$	$455\frac{5}{8}$	
B, 5#		$256\frac{37}{128}$		$284\frac{49}{64}$	$303\frac{3}{4}$		$341\frac{23}{32}$		$379\frac{11}{16}$		$427\frac{19}{128}$	$455\frac{5}{8}$	
E, 4#		$253\frac{1}{8}$		$284\frac{49}{64}$	$303\frac{3}{4}$		$341\frac{23}{32}$		$379\frac{11}{16}$	405		$455\frac{5}{8}$	
A, 3#		$253\frac{1}{8}$	270		$303\frac{3}{4}$		$337\frac{1}{2}$		$379\frac{11}{16}$	405		$455\frac{5}{8}$	
D, 2#		$253\frac{1}{8}$	270		$303\frac{3}{4}$		$337\frac{1}{2}$	360		405		450	
G, 1#	240		270		300		$337\frac{1}{2}$	360		405		450	480
C	240		270		300	320		360		400		450	480
F, 1♭	240		$266\frac{2}{3}$		300	320		360		400	$426\frac{2}{3}$		480
B♭, 2♭	240		$266\frac{2}{3}$	$284\frac{4}{9}$		320		$355\frac{5}{9}$		400	$426\frac{2}{3}$		480
E♭, 3♭	$237\frac{1}{27}$		$266\frac{2}{3}$	$284\frac{4}{9}$		320		$355\frac{5}{9}$	$379\frac{7}{27}$		$426\frac{2}{3}$		$474\frac{2}{27}$
A♭, 4♭	$237\frac{1}{27}$	$252\frac{68}{81}$		$284\frac{4}{9}$		$316\frac{4}{81}$		$355\frac{5}{9}$	$379\frac{7}{27}$		$426\frac{2}{3}$		$474\frac{2}{27}$
D♭, 5♭	$237\frac{1}{27}$	$252\frac{68}{81}$		$284\frac{4}{9}$		$316\frac{4}{81}$	$337\frac{29}{243}$		$379\frac{7}{27}$		$421\frac{97}{243}$		$474\frac{2}{27}$
G♭, 6♭		$252\frac{68}{81}$		$280\frac{680}{729}$		$316\frac{4}{81}$	$337\frac{29}{243}$		$379\frac{7}{27}$		$421\frac{97}{243}$	$449\frac{359}{729}$	
	C	D♭	D	E♭	E	F	G♭	G	A♭	A	B♭	C♭	C'

Just scales from 6 flats to 6 sharps showing frequencies.

It is clear that just temperament is tricky. In the case of Pythagorean temperament, the scales from three flats to two sharps were all perfect with a 12-note keyboard. But in just intonation only one scale (C) is perfect. Alternative solutions were thus sought. One which was used widely until the middle of the nineteenth century was the meantone system. This in turn was superseded by the system of equal temperament used today. Before looking at these we will introduce minor scales which add another twist to the story.

Minor scales

SO FAR we have only considered major scales. But there are also minor scales with rather different ratios. We introduce minor scales by considering thirds.

In the major scale the interval CG (a fifth) is divided into two thirds by E. CE is a major third which in just temperament has ratio 5/4. And since CG is 3/2 it follows that EG has ratio 6/5. This is known as a minor third. Recall that a major tone is 9/8 whereas a minor tone is 10/9.

The minor third has a melancholy sound compared to a major third. In a major scale the major third precedes the minor third and somewhat obscures it. In a minor scale the fifth CG is subdivided with the minor third preceding the major third. A minor third comprises three semitones compared with the four semitones of a major third. So the minor scale starts

CDE♭FG

and the melancholy flavour is obvious.

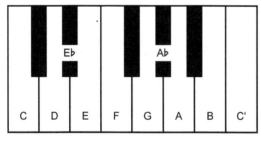

The frequencies and ratios in the Harmonic Minor scale.

Harmonic Minor.

The second part of the scale could be structured in the same way. Take the fifth from F to C' and note that in the major scale it is subdivided by A into a major third followed by a minor third. In a minor scale it would be expected that this fifth would be subdivided into a minor third followed by a major third. In other words that A would be replaced by A♭ such that the ratio FA♭ is 6/5. And indeed A♭ figures in the minor scale starting at C. Note that CA♭ (the just minor sixth) is 8/5.

However, life is made complicated by the fact that there are two forms of minor scale in which the part from F to C' varies. First there is the Harmonic Minor which starting from C is

    CDE♭FGA♭BC   Harmonic Minor

Note the huge interval A♭B between two successive notes. This is three semitones and has ratio 75/64.

This huge interval is avoided in the other form of minor scale known as the Melodic Minor. But this strangely is different when ascending and descending. When ascending, the melodic minor is like the major scale from F onwards and does not use A♭.

But descending it uses both A♭ and B♭ thus

    CDE♭FGABC   Melodic Minor ascending

    CDE♭FGA♭B♭C   Melodic Minor descending

So the melodic scale has the familiar intervals of the major scale but in a different order.

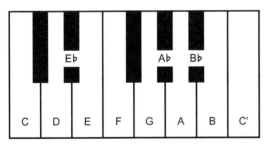

C	D	E♭	E	F	G	A♭	A	B♭	B	C'	
		288				384	432				
240	270		(300)	320	360		400		450	480	
		6/5				8/5	9/5				
1/1	9/8		(5/4)	4/3	3/2		5/3		15/8	2/1	
	9/8	16/15	10/9	9/8		10/9		9/8		16/15	Melodic Minor ascending.
	9/8	16/15	10/9	9/8		16/15		9/8		10/9	Melodic Minor descending.

The frequencies and ratios in the Melodic Minor scales.

note	minor	Pythagorean	error in cents
C	240	240	0
D	270	270	0
E♭	288	$284^4/_9$	−22
F	320	320	0
G	360	360	0
A♭	384	$379^7/_{27}$	−22
A (asc)	400	405	+22
B♭ (dsc)	432	$426^2/_3$	−22
B	450	$455^5/_8$	+22
C'	480	480	0

The scale of C minor and Pythagorean temperament.

The frequencies of the just minor scale starting from C are shown in the table above. It is interesting to consider how well the Pythagorean temperament matches the just minor scales.

If we consider C minor then we find that C, D, F, G, C' are perfect. But there are discrepancies in the others. The table above also shows the Pythagorean frequencies and the error in cents. We see that E♭, A♭, and B in the Harmonic minor scale are all in error by a syntonic comma. In the case of the Melodic minor scale we find that A and B♭ are also in error by a syntonic comma. Curiously, if we had compromised on tuning our Pythagorean piano for G#/A♭ midway between them, then the error in A♭ would be much less.

Similarly, we can consider the C minor scale on a piano tuned in just temperament with the black keys C#, E♭, F#, and B♭ correct and the compromise on G#/A♭. We find that E♭, A♭, and B♭ are still wrong for the minor scale but A and B are now correct.

So the general conclusion is that the Pythagorean temperament is very unsatisfactory for the minor scales. The justly tuned keyboard is a little better but not much.

We finish by noting that the chord comprising the three notes CEG is known as a major triad whereas the chord CE♭G is known as a minor triad. These are illustrated below together with the Harmonic minor scale of C. We will encounter the minor scales again and lots of triads in Appendix J.

major triad   minor triad   The Harmonic minor scale.

Meantone temperament

THE GENERAL IDEA was to remove the difference between the major tone of 9/8 and the minor tone of 10/9. In fact the major third of 5/4 was kept exact and so the tones became $\sqrt{(5/4)} = 1.118$ .... This also applied to the tones FG, GA, and AB; the two semitones EF and BC' then shared the rest of the ratio of 2 for the whole octave and so became about 1.070.

The result is shown below which compares the values with the Pythagorean and just scales. The meantone fifth is a bit flat but is a better approximation to a just fifth (error is about 5 cents) than the Pythagorean third is to a just third (22 cents). The semitone is a bit larger than half a tone in the meantone scale whereas it was a lot smaller than half a tone in the Pythagorean scale.

We can now develop the black notes by transposition just as we did before. We only require one new note (the black one) because all the tones are the same.

Thus we make F# a tone above E giving 335.4 and C# a tone above B giving 250.8. And we make B♭ a tone below C giving 429.3 and E♭ a tone below F giving 287.1. Again we find that G# should be a tone above F# giving exactly 375, whereas A♭ should be a tone below B♭ giving exactly 384. Again we find that A♭ and G# are not the same. Note that G# is a whole number because it is two tones above E and we have chosen the system so that E is exactly 300 and two tones are 5/4. A similar remark applies to A♭ which is two tones below C'.

Interestingly, whereas G# was above A♭ in the Pythagorean system, it is below in the meantone system. The difference between G# and A♭ is exactly 128/125 – this ratio is called the Great Diesis and is 42 cents. The difference in the Pythagorean system was the Pythagorean Comma of 24 cents. So the meantone system has a larger difference here and perhaps is a hint that it is going to get into trouble when we stray far from C.

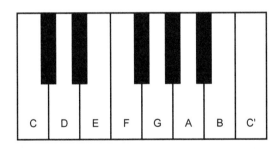

Introducing the meantone system.

C	D	E	F	G	A	B	C'	
240	270	303¾	320	360	405	455⅝	480	Pythagorean.
240	270	300	320	360	400	450	480	Just intonation.
240	268.3	300	321.0	358.9	401.3	448.6	480	Meantone.
	1.118	1.118	1.070	1.118	1.118	1.118	1.070	

The table below shows these values and for comparison includes the Pythagorean and just values and the difference between the meantone and just values in cents.

If we tune our piano so that F#, C#, B♭, and D♭ are correct and compromise for G#/A♭ then we find that the meantone system works very well for keys close to C. The acid test is the fifths. We know that CG is slightly flat but it is OK. The ratio is 1.495 whereas it should be 1.5 of course – an error of about 5 cents. But the G#/A♭ problem is nasty and as a consequence the fifths from C# to G# and from A♭ to E♭ are in error by 15 cents assuming that we have compromised on the tuning of the one key G#/A♭. Moreover, some thirds are bad. Thus CE is exactly 1.25 as arranged but C#F and F#B♭ are 1.280 (an error equal to the Great Diesis of 128/125 or 42 cents). These intervals were called wolves because chords using them howled with discordance.

Minor scales close to C are not too bad. For C minor, CE♭ is 1.196 as opposed to the ideal 1.2 – an error of about 6 cents. Moreover, A♭ is exactly correct for the minor sixth CA♭ of ratio 8/5 although if we have compromised on G#/A♭ then there will be an error of 21 cents (half a Great Diesis).

So the meantone system works quite well for keys close to C. Of course it is not so good as the just system in C itself but as we have seen the just system on a traditional keyboard breaks down immediately we depart from C.

	meantone	Pythagorean	just	m/j cents
C	240	240	240	0
C#	250.774	256.289	253.125	16
D♭	256.792	252.840	252.840	27
D	268.328	270	270	11
D#	280.374	288.325	284.766	27
E♭	287.103	284.444	284.444	16
E	300	303.75	300	0
F	320.990	320	320	5
F#	335.410	341.718	337.5	11
G♭	343.460	337.119	337.119	32
G	358.878	360	360	5
G#	375	384.434	379.688	22
A♭	384	379.259	379.259	21
A	401.238	405	400	5
A#	419.263	432.488	427.148	32
B♭	429.325	426.667	426.667	11
B	448.598	455.625	450	5
C'	480	480	480	0

Meantone versus Pythagorean and just temperaments.

There were variations on the meantone system. Indeed the Pythagorean system can be thought of as a meantone system because the tones are all equal. In the Pythagorean system, F and G are perfect, whereas E is in error by a syntonic comma (we will just say comma in future) and in the meantone system E is perfect, but F and G are in error by a quarter of a comma. To see why this is exactly the case observe that four intervals of a fifth from C bring us to E thus

$$C \to G \to D \to A \to E$$

So by removing the error of a comma in E it follows that the fifths must all be smaller by a quarter of a comma.

We can consider a variation whereby the tone is chosen so that the errors in E and G are the same. Suppose that the intervals of a tone and semitone are ratios of T and S respectively. Then the errors in E and G will be (as ratios)

$T^2 \div 5/4$ error in E

$T^3 S \div 3/2$ error in G

If these are to be equal and opposite then their product must be 1 so we have

$$T^5 S = 15/8$$

but since an octave comprises five tones and two semitones we also have

$$T^5 S^2 = 2$$

From these two equations we deduce that S is 16/15, the just semitone! With this version, E and G are both in error by a fifth of a comma, so this version of meantone is sometimes called 1/5-comma meantone. Similarly, the original meantone is 1/4-comma meantone (G being in error by 1/4-comma) and Pythagorean is 0-comma meantone (G being exact). The 1/5-comma meantone is a bit better than 1/4-comma but scales inevitably get bad if we stray far from C.

An interesting point is that in any meantone system the error in E is always the sum of the errors in A and G paying due attention to the signs of the errors. Thus if E is correct then A and G have equal errors but in opposite directions (1/4-comma), whereas if G is correct then A has the same error as E and in the same direction (a comma).

Another aspect of the meantone and Pythagorean system is that the flats and sharps are unequal. In the Pythagorean system G# is above A♭ whereas in 1/4-comma meantone their order is reversed. We might therefore consider a meantone system whereby G# and A♭ are the same. A little thought reveals that this means that all semitones must be exactly half a tone so that $T = S^2$. This brings us immediately to the equitempered system.

The meantone system (1/4-comma) was widely used from the late seventeenth century until the middle of the nineteenth century. But composers then got more adventurous and did not like being confined to scales close to C. The desire for more freedom led to the equitempered system.

Equal temperament

IN THE NINETEENTH CENTURY it was generally realised that perhaps the best overall solution was to divide the 12 semitones between C and C' equally and to make them each equal to the 12th root of 2. And that is how a piano is usually tuned today.

None of the intervals (other than the octave) is perfect but the errors are fairly uniformly distributed and all keys are equal. So the interval from C to G has ratio of $2^{7/12}$ which is 1.4983... whereas ideally it should be 1.5. The error is very close indeed to 1/11 of a comma (strictly 1/11.000651...). Similarly the interval from C to E has ratio $2^{4/12}$ which is 1.2599..., whereas ideally it should be 1.25. The error is some 0.63638... of a comma, very close to 7/11 of a comma. A significant error but not as bad as the error of a whole comma which turns up in other systems.

As a consequence none of the scales is perfect but none is really awful. The diagram below shows the frequencies for the various scales.

For those that were the same in the Pythagorean and just scales, the equitempered one differs but little. For those that were different in the Pythagorean and just scales the equitempered scale strikes a reasonable compromise value. Note that the equitempered scale is better than meantone regarding fifths but not so good for thirds. Indeed thirds are its weakness.

The table opposite shows the frequencies of all notes and gives the difference between the just and equitempered values in cents. The worst error is in A (major sixth) at 16 cents, followed by E at 14 cents and B at 12 cents.

We should also have a quick look at the minor scales. The ideal E♭ (so that CE♭ is 6/5) is 288 and this has an error of 16 cents (the same as A which is as

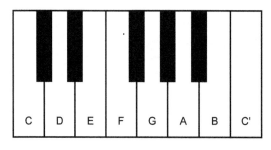

240	270	303¾	320	360	405	455⅝	480	Pythagorean.
240	270	300	320	360	400	450	480	Just intonation.
240	268.3	300	321.0	358.9	401.3	448.6	480	Meantone.
240	269.4	302.4	320.4	359.6	403.6	453.1	480	Equitempered.

The frequencies in the various major scales.

expected because a minor third is simply the inverse of a major sixth). The ideal A♭ (so that CA♭ is 8/5) is 384 and this has an error of 14 cents (the same as E which is as expected because a minor sixth is the inverse of a major third). So at least the minor scales are no worse than the major scales.

An interesting comparison of all the various scales is provided by the diagram overleaf. The vertical scale gives the frequency in cents and the horizontal axis describes the meantone error in G. At the left extreme is the Pythagorean scale (which has exact G) where the sharps (such as C#) are sharper than the corresponding flats (such as D♭), and at the right extreme is the 1/4-comma meantone (which has exact E but a 1/4-comma error in G) where the opposite holds. The lines for the various sharps and flats cross over at the equitempered scale. The short dashes at the sides are the multiples of 100 cents and thus mark the equitempered semitones. The horizontal red dotted lines represent the just notes in the major scale (D, E, F, G, A, B) whereas the blue dotted lines represent the extra notes in the just minor scales (E♭, A♭, B♭).

Note that G and F are acceptably good in all cases. E is awful in the Pythagorean scale but perfect in 1/4-comma meantone (A♭ is similar to E). A and E♭ are also bad in the Pythagorean scale but good in 1/4-comma meantone. And B is perfect in 1/5-comma meantone. Furthermore, D is perfect in the Pythagorean scale (but D is relatively unimportant). It is also clear that the equitempered scale is not ideal – E, A, and B are significantly in error.

	equi	just	e/j cents
C	240	240	0
C#	254.271	253.125	8
D♭	254.271	252.840	10
D	269.391	270	4
D#	285.410	284.766	4
E♭	285.410	284.444	6
E	302.381	300	14
F	320.362	320	2
F#	339.411	337.5	10
G♭	339.411	337.119	12
G	359.594	360	2
G#	380.976	379.688	6
A♭	380.976	379.259	8
A	403.630	400	16
A#	427.631	427.148	2
B♭	427.631	426.667	4
B	453.060	450	12
C'	480	480	0

Equitempered versus just temperament.

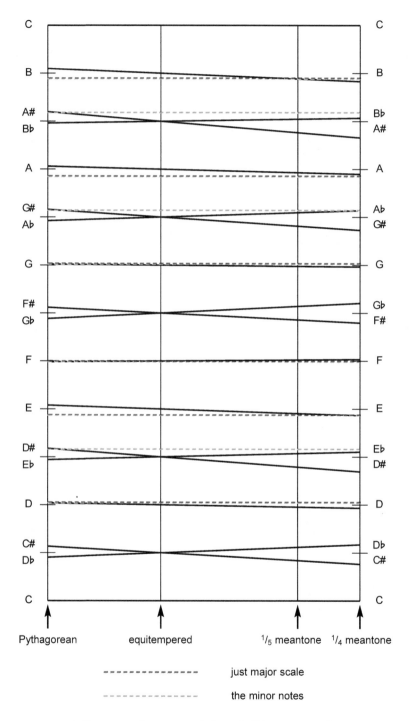

A diagrammatic representation of all temperaments.

Nevertheless, one important practical advantage of the equitempered scale is that the conflict between flats and sharps which was a real problem with the Pythagorean and meantone systems is not present. Another huge advantage of the equitempered scale is that all keys are equally good (or bad) and composers are free to stray as far from C as they like. So, although clearly not ideal, the 12-note equitempered scale is now the accepted standard. But although it was introduced in the eighteenth century (and even earlier for some fretted instruments such as the viol), it was not universally accepted for a long time. Organs in particular remained tuned in the meantone scale for many years.

Even though the equitempered scale is the norm for keyboard instruments, string instruments can be tuned justly for the particular scale being used and indeed often are unless being accompanied by a piano. Equally, choirs can use the just scales as appropriate.

Although we are so used to the equitempered scales, it is said that to hear an organ tuned to a just scale and to enjoy the full majesty of the concordance of a major chord reveals what is missed by our compromise.

We finish this discussion with a table showing the pure intervals that we have encountered and how they fare against the various temperaments. At a glance this reveals that the equitempered scale lies in the middle ground between the other scales.

When discussing the just scale in terms of multiples of some fundamental frequency, we noted that the multiples $7f$ and $11f$ were not accommodated. This remains so for the true 7th in the equitempered scale. We need a note of 969 cents above C but this is not closely matched by any of the equitempered notes. And similarly the true 11th is 1751 cents above C and thus midway between equitempered F and F#.

interval	Pythagorean	just	meantone	equitempered
semitone FG	90	112	117	100
minor tone (10/9) DE	204	182	193	200
major tone (9/8) CD	204	204	193	200
minor third (6/5) CE♭	294	316	310	300
major third (5/4) CE	408	386	386	400
fourth (4/3) CF	498	498	503	500
fifth (3/2) CG	702	702	697	700
minor sixth (8/5) CA♭	792	814	814	800
major sixth (5/3) CA	906	884	890	900
true seventh (7/4)		969		??
minor seventh (9/5) CB♭	996	1017	1007	1000
octave	1200	1200	1200	1200

The pure intervals showing how they are met by the various temperaments.

Other arrangements

MANY OTHER ARRANGEMENTS have been devised especially regarding the subdivision of major and minor tones. It is clear that there was a need to have a single key for C#/D♭ etc. and so it remained to decide in the just scale how to subdivide a major tone of 9/8 and a minor tone of 10/9.

For some reason there was a very strong desire to stick to whole numbers and so the meantone style of using ratios such as √(9/8) was considered unacceptable.

For example, the English cleric Thomas Salmon (1648–1706) divided 9/8 into 18/17 and 17/16, which can be neatly expressed as 18:17:16 and he divided 10/9 into 20:19:18 – in both cases the smaller subdivision came first. This results in C#/D♭ being 254.118 whereas the just values calculated earlier (for scales near C) are $253\frac{1}{8}$ and $252\frac{68}{81}$. So the Salmon system gives a value which is about 8 cents too high.

Other ways of splitting a major tone of 9/8 are into an interval of 27/25 (Great Limma) and one of 25/24 (small semitone) or into 16/15 (diatonic semitone) and 135/128 (Larger Limma).

And a minor tone of 10/9 can be split into 16/15 (diatonic semitone) and 25/24 (small semitone). Note further that 25/24 (small semitone) can be split further into 250/243 and 81/80 (syntonic comma).

We will encounter some of these amazing ratios when we look at Mersenne's keyboards in Appendix J.

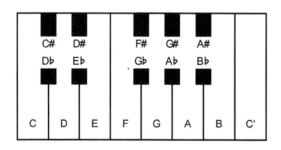

Salmon's subdivision of the just scale.

Ratios of all semitones.
Ratios of white notes.

Dividing the octave

THE REASON for the acceptance of the equitempered scale is that $2^{7/12}$ is a good approximation to 3/2 for a perfect fifth and so on. The question naturally arises as to whether other equal subdivisions of an octave would give more accurate approximations to the important intervals.

In particular, although the fifth is accurate to about 1 part in 1000, the major third is only accurate to 1 part in 120. And of course it was the rough nature of the major third in the Pythagorean scale that triggered the introduction of just intonation and then meantone. So maybe we should seek a scale that gives a better major third as well as having a good perfect fifth.

As just noted, the reason that the equitempered 12-note octave gives a good approximation to a perfect fifth is that $2^{7/12}$ is very close to 3/2. In other words

$\log_2 3/2$ is close to 7/12

So we seek other rational approximations to $\log_2 3/2$ for a good perfect fifth. Similarly, we can seek other rational approximations to $\log_2 5/4$ for a better major third and so on. The usual way to find rational approximations is to express a value as a continued fraction and then successive convergents give better and better approximations as explained in the lecture on Fractions. In the case of $\log_2 3/2$ we get

$\log_2 3/2 = [1, 1, 2, 2, 3, 1, 5, ...]$

and the convergents are 1/1, 1/2, 3/5, 7/12, and so on.

So we get the following approximations

7/12, 24/41, 31/53, 179/306	perfect fifth
1/3 (= 4/12), 9/28, 19/59, 47/146	major third
1/6 (= 2/12), 8/47, 9/53, 26/153	major tone

The numerator is the number of equal notes in an octave (or a multiple thereof). The approximation 7/12 for a fifth corresponds to the normal 12-note scale and confirms that the approximation to a fifth is good. 4/12 confirms that the major third is reasonable and 2/12 confirms the major tone.

It might remembered that if an approximation has small numbers compared to the next one then it is particularly good. So 31/53 will be excellent for a fifth. Moreover, 9/53 is also excellent for a major tone. So it looks as if a 53-note equitempered scale might be excellent.

Two other intervals are worth considering. One is the seventh with ratio 7/4 (and so the seventh harmonic of the low C mentioned earlier) which would be 420 on our scale – we have to use B♭ at $426^{2}/_{3}$ which is not at all good. The other is the minor third of ratio 6/5. The scale of C minor has E♭ instead of E and so

E should be 288 for this purpose. It is interesting to look at the approximations for these two intervals. We get

1/4, 5/19, 111/422 minor third

4/5, 21/26, 88/109 seventh

The minor third is reasonably well placed in the 12-note scale since 1/4 is 3/12. But note the amazing approximation 5/19 followed by the huge 111/422. This means that a 19-note scale would have fabulous minor thirds. The seventh looks a bit of a puzzle.

It is clear that the 53-note scale is worth exploring. We also look at the 19-note scale (which has merits for the minor third) and at the 31-note scale which has also been proposed in the past. The small numbers before the frequencies in the table below give the note position in the scale. Thus in the 12-note scale, E is the fourth note above C whereas in the 53-note scale E is the seventeenth note.

It is clear that the 53-note scale is far superior to the 12-note scale in every respect. The 19-note scale is superior to the 53-note scale just for the notes E♭ and A and is not markedly better than the 12-note scale generally – it is no doubt good for minor keys. The 31-note scale is excellent for E as expected and for some other notes such as E♭, A♭, A, and B♭ it is better than the 12-note scale, but for the all important F and G it is not so good as the 12-note scale.

The 19- and 31-note scales do not seem of much overall value. But the 53-note scale is obviously very good indeed. Most notes are stunning and it even has a reasonable note for the poor seventh. The errors in cents are given in the table opposite for the 12-note and 53-note scales.

The worst offenders in the 12-note major scale are E and A and these are ten times better in the 53-note scales. The additional offenders in the minor scale

note		just	12-note	19-note	31-note	53-note
C	1/1	240	0 240	0 240	0 240	0 240
D	9/8	270	2 269.39	3 267.76	5 268.39	9 269.98
E♭	6/5	288	3 285.41	5 288.03	8 287.01	14 288.22
E	5/4	300	4 302.38	6 298.73	10 300.14	17 299.76
F	4/3	320	5 320.36	8 321.34	13 320.96	22 320.01
G	3/2	360	7 359.59	11 358.50	18 358.92	31 359.99
A♭	8/5	384	8 380.98	13 385.64	21 383.83	36 384.31
A	5/3	400	9 403.63	14 399.97	23 401.38	39 399.69
7th	7/4	420	10 427.63	15 414.83	25 419.74	43 421.16
B♭	9/5	432	10 427.63	16 430.24	26 429.23	45 432.32
C'	2/1	480	12 480	19 480	31 480	53 480

Several equitempered scales showing frequencies.

note		just	12-note		error	53-note		error
C	1/1	0.00	0	0.0	0.00	0	0.0	0.00
D	9/8	203.91	2	200.0	3.91	9	203.77	0.14
E♭	6/5	315.64	3	300.0	15.64	14	316.98	1.34
E	5/4	386.31	4	400.0	13.69	17	384.91	1.40
F	4/3	498.05	5	500.0	1.95	22	498.11	0.06
G	3/2	701.95	7	700.0	1.95	31	701.89	0.06
A♭	8/5	813.69	8	800.0	13.69	36	815.09	1.40
A	5/3	884.36	9	900.0	15.64	39	883.02	1.34
7th	7/4	968.83	10	1000.0	31.18	43	973.59	4.76
B♭	9/5	1017.60	10	1000.0	17.60	45	1018.87	1.27
C'	2/1	1200.00	12	1200.0	0.00	53	1200.00	0.00

The values and errors in cents for the 12-note and 53-note equitempered scales.

(which are E♭, A♭, and B♭) are also ten times better. But the poor seventh, although greatly improved is still in error by nearly 5 cents.

In around 1876, the English musicologist Robert H M Bosanquet (1841–1912) actually constructed a harmonium with 53 notes to the octave.

However, for perfection and especially for the seventh there was no alternative to the construction of multiple keys for just intonation. The American engineer Henry Ward Poole (1825–1890) constructed an organ which covered all keys in just intonation from five flats to five sharps; it seems to have had 100 notes to the octave.

We finish with a few words about chords. As we have noted, CE is a major tone and EG is a minor tone. The chord CEG is a major third and the frequencies of the notes are in the ratio 4–5–6, and adding C' gives the major common chord shown below with ratios 4–5–6–8. How nice it would be to have a proper seventh with frequencies 4–5–6–7. But alas the best we can do in 12-note equitempered scale is to use B♭ which is in error by 32 cents.

But it is clear that the 12-note equitempered scale is here to stay. Really the only thing that is wrong with it is the chord of the seventh CEGB♭ shown below. However, in practice this chord sounds quite reasonable – perhaps the richness of the harmonics hides the inaccuracy.

And on that note we conclude this curious topic.

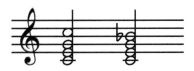

Major common chord   Chord with seventh

Further reading

A CLASSIC WORK on this topic is *On The Sensations of Tone* by Hermann Helmholtz. The final German edition was in 1877. It was translated shortly afterwards by Alexander Ellis. It is now available in Dover. My copy is dated 1954. This is a large book and is hard going. It covers in enormous detail many aspects of harmony and the various organs and scales that have been devised. It could drive one insane. A more accessible discussion will be found in Chapter 1 of *Music and Mathematics* edited by Fauvel, Flood, and Wilson.

Detailed discussions regarding scales will be found in *Tuning and Temperament* by J Murray Barbour and in *Lutes and Viols and Temperaments* by Mark Lindley.

See also Appendix J on Mersenne's keyboards which takes an alternative approach to understanding the construction of multiple scales.

10 Finale

THIS FINAL LECTURE looks at a few topics of a varied nature. We start by considering the use of prime numbers in cryptography. We also look at the curious subject of animal gaits and conclude by considering two classic puzzles – the Tower of Hanoi and Chinese Rings.

The RSA algorithm

UNTIL RECENTLY it was thought that there were no practical applications for the use of prime numbers. However, the advent of the internet has created a need for the encryption of messages.

Generally speaking a message is encrypted and decrypted with the aid of some sort of key. For example, an ancient technique uses a simple key word. Suppose the key word is Numbers. We then write the word Numbers followed by the remaining letters of the alphabet in order and underneath write the full alphabet in order thus

```
n u m b e r s a c d f g h i j k l o p q t v w x y z
a b c d e f g h i j k l m n o p q r s t u v w x y z
```

In order to encrypt a message such as "Send more money", we take each letter, find it in the first line and replace it by the letter underneath it. So "Send more money" becomes "Geaj crfe craey".

The receiver of the message does the reverse, finds each letter in the bottom line and replaces it by the one above it. Clearly this can only be done if both sender and receiver know the key. If a spy discovers the key that the sender has used then he can decrypt the message.

So the key has to be kept secret. Moreover, the key has to be changed frequently otherwise it is easy to decode messages based on the frequency of letters and so on if several messages with the same key are intercepted. Typically, both sender and receiver have secret code books giving the key to be used on each day. These are, of course, vulnerable to theft.

This problem was overcome by introducing asymmetry whereby the decryption process was not simply the reverse of encryption. Thus the key for encryption could be public and yet knowledge of it would not be sufficient to do the corresponding decryption

An amazing algorithm based on the use of prime numbers was invented by Ronald Rivest, Adi Shamir, and Leonard Adleman in 1977 and so known as the RSA algorithm. The main point is that it is very easy to multiply two large prime numbers, p and q, together to give n say. But it is quite difficult to start from n and discover its prime factors. So the receiver R (Roger perhaps) thinks of two large prime numbers but keeps their identity to himself. He multiplies them together to give n and makes this public and says to the world "Use the key n to send me secret messages". So any sender S (Sarah or Susan or Simon) can send messages to Roger. However, the decoding process requires knowledge not just of the key n but actually of the factors p and q. Roger knows these and accordingly can decode the messages. But the Spy does not know these and cannot read the messages.

The mechanics is as follows. The public key actually comprises two numbers, the number n which is the product of two primes p and q and a second number e which is chosen so that e is less than and relatively prime to $m = (p-1) \times (q-1)$. Remember that relatively prime means that e and m have no common factors so that $\gcd(e, m) = 1$.

An important property of the fact that e and m have no common factors is that there is a unique number d (less than m) such that

$e \times d \equiv 1 \bmod m$

For example if m is 10 and e is 7 then the numbers $e, 2e, 3e, 4e, ..., 9e$ (mod 10) are all distinct and less than 10. They are the last digits of the multiples of 7 (in base 10) and are 7, 4, 1, 8, 5, 2, 9, 6, 3. And since they are distinct one of them must be a 1. Thus we see that $7 \times 3 \equiv 1 \bmod 10$. In a sense 3 is the inverse of 7.

This distinctness is fairly easy to prove. Suppose it were not true and that $a \times e$ and $b \times e$ were the same, where both a and b are positive and different and less than m. This would mean that $(a-b) \times e \equiv 0 \bmod m$. In other words $(a-b) \times e$ is a multiple of m say $(a-b) \times e = k \times m$. Now both sides of this must have the same prime factors, and moreover we know that m and e have no factors in common. It follows that all the factors of m must be factors of $a-b$ which means that $a-b$ is a multiple of m. However, we also know that both a and b are less than m and this can only be possible if a and b are the same which is a contradiction. It then follows that $a \times e$ and $b \times e$ cannot have been the same. QED.

We are now in a position to do the RSA algorithm. The values e and n are made public but p and q and also d remain secret. Note that we are using e for encrypt and d for decrypt.

Individual letters of the message are converted into a standard numeric value (the Latin-1 value perhaps, so that A is 65, B is 66 and so on).

A letter of value v is then encrypted into the code value c using the formula

$c = v^e \bmod n$

The code value c then comprises the message. Decryption uses the secret number d according to a similar formula thus

$v = c^d \bmod n$

and gives the original value.

An example might clarify this. Suppose p and q are the primes 13 and 23. Then we have

$n = p \times q = 13 \times 23 = 299$

$m = (p-1) \times (q-1) = 12 \times 22 = 264$

Now we choose e relatively prime to 264, we can take $e = 5$. Finally, we need to find the unique d such that $e \times d \equiv 1 \bmod 264$. It is in fact 53 since $5 \times 53 = 265$. So we have

$e = 5$

$d = 53$

The public key is now the two numbers $e = 5$ and $n = 299$.

To encrypt the letter B = 66 we compute $66^e \bmod n$ which is

$66^5 \bmod 299$

We do this by calculating 66, 66^2, 66^4 and noting that $66^5 = 66^{1+4} = 66^1 \times 66^4$. At each stage we can take the remainder on division by 299 so the numbers do not get too large. We have

$66^2 = 4356 \equiv 170 \bmod 299$

$66^4 \equiv 170^2 \equiv 28900 \equiv 196 \bmod 299$

$66^5 = 66 \times 66^4 \equiv 66 \times 196 \equiv 79 \bmod 299$

and 79 is the encrypted message value.

The receiver reverses the process by computing $79^{53} \bmod 299$ in a similar manner. Now $53 = 32 + 16 + 4 + 1$ (we simply convert 53 to binary). So we get

$79^2 = 6241 \equiv 261 \bmod 299$

$79^4 \equiv 261^2 \equiv 68121 \equiv 248 \bmod 299$

$79^8 \equiv 248^2 \equiv 61504 \equiv 209 \bmod 299$

$79^{16} \equiv 209^2 \equiv 43681 \equiv 27 \bmod 299$

$79^{32} \equiv 27^2 \equiv 729 \equiv 131 \bmod 299$

$79^{53} = 79 \times 79^4 \times 79^{16} \times 79^{32} \equiv 79 \times 248 \times 27 \times 131 \equiv 66 \bmod 299$

and amazingly we have recovered the 66. So it works.

In this example, the factorization of 265 is quite easy but if two very large primes were taken then the factorization could be prohibitively difficult. The calculations involved in doing the powers may seem tedious but can easily be automated and done quickly on a computer.

p	q	m	m+1	factors	e	d
11	13	120	121	11^2	11	11
11	17	160	161	7.23	7	23
11	19	180	181	prime	7	103
13	17	192	193	prime	17	113
13	19	216	217	7.31	7	31
13	23	264	265	5.53	5	53
17	19	288	289	17^2	17	17
17	23	352	353	prime	19	315
19	23	396	397	prime	23	155

Some pairs of primes *p* and *q* and exponents *e* and *d* for the RSA algorithm.

The table above shows various pairs of small primes and some corresponding values for *e* and *d*.

If $m+1$ is not prime then we can simply choose a pair of factors for *e* and *d*. If it is prime then we just choose a favourite number for *e* and then have to find *d* as explained later.

Note that if the two primes are adjacent odd numbers such as $p = 11$ and $q = 13$, then $m+1$ is always the square of the smaller and we could choose *e* and *d* to be the same. However, this might be slightly insecure because it turns out that some numbers are actually unchanged when encoded (namely $q-1$, q, and $q+1$ which in the case of this example are 12, 13, and 14). Of course we don't have to use actual factors of $m+1$ and in the case of 11 and 13 we can take $e = 23$ and $d = 47$. Incidentally 0 and 1 always remain the same when encoded.

A nice example is with $p = 13$ and $q = 19$ which gives $m+1 = 217$ which is the product of the Mersenne primes 7 and 31. So we can take $e = 7$ and $d = 31$. It is unclear whether this has any special properties.

There are two loose ends. One is the problem of finding a pair for *e* and *d*. If we choose an *e* without thinking then the corresponding *d* can be awkward to find. As we have just seen, if $m+1$ is not prime then a pair of factors of $m+1$ will do. Thus, in the example above $5 \times 53 = 265 = m+1$. But if $m+1$ is large finding a pair of factors might be tedious and if $m+1$ is actually prime, almost impossible.

The other loose end is why does the algorithm work? In other words why does the decryption process using *d* reverse the encryption process using *e*? In encryption we raise the value to the power *e*. For decryption we need to reverse this process and take the eth root. The secret lies in the fact that $e \times d \equiv 1 \bmod m$ so that *d* is a sort of inverse of *e* modulo *m*. From this it can be shown that taking the eth root is the same as raising the value to the power *d*. But this glib explanation needs some justification. We will now look at these two loose ends and then return to the RSA algorithm.

Linear congruences

IN ORDER to discover how to find a value for d such that $e \times d \equiv 1 \bmod m$ we need to briefly introduce Diophantine equations. These are named after Diophantus, the Greek mathematician and philosopher, who lived in Alexandria around 250. Consider the following equation

$$ax + by = c$$

where a, b, and c are integers. The question is: can we find integers x and y that satisfy this equation? An example might be

$$15x + 55y = 325$$

and one solution is $x = 7$ and $y = 4$. Another is $x = 18$ and $y = 1$. Sometimes such equations have no solution at all. For example

$$2x + 4y = 17$$

has no solutions. This is obvious since whatever values we give to x and y, the left hand side is bound to be even and so cannot equal the right hand side which is odd.

In fact the equation only has a solution if the gcd of a and b is a factor of c. Remember that the gcd is the greatest common divisor or the largest number that divides into both a and b.

In the case of 15 and 55, their gcd is 5 and since 5 is a factor of 325 then solutions to the first equation do exist. On the other hand, the gcd of 2 and 4 is 2 and 2 is not a factor of 17 and so the second equation has no solutions.

If it exists, a solution can be found by using Euclid's algorithm which we met in the lecture on Primes. We have

$$55 = 3 \times 15 + 10 \qquad \text{line 1}$$
$$15 = 1 \times 10 + 5 \qquad \text{line 2}$$
$$10 = 2 \times 5 + 0 \qquad \text{the gcd is 5 because the remainder is 0}$$

When computing the gcd we are only interested in the remainder and the answer is the last divisor when the remainder becomes 0 which is 5 in this case. But in this application we need to take account of both quotient and remainder.

We now work backwards

$$5 = 15 - 1 \times 10 \qquad \text{using line 2 of gcd calculation}$$
$$= 15 - 1 \times (55 - 3 \times 15) \qquad \text{using line 1}$$
$$= 4 \times 15 - 55 \qquad \text{gathering the terms in 15}$$

We now rearrange and get

$$15 \times 4 - 55 \times 1 = 5$$

and then we multiply this by $325 \div 5 = 65$ (remember that 5 is the gcd) and get

$$15 \times 260 - 55 \times 65 = 325$$

So we now have a solution, namely $x = 260$ and $y = -65$. Now $11 \times 5 = 55$ and $3 \times 5 = 15$. This means that if we increase x by 11 and decrease y by 3 (or decrease x by 11 and increase y by 3) then the left hand side will stay the same and so we then have another solution. In fact if we decrease x by 23×11 and increase y by 23×3 then we get the solution $x = 260 - 253 = 7$ and $y = -65 + 69 = 4$ that was mentioned earlier. Increasing x again by 11 and decreasing y by 3 gives the other solution mentioned earlier, namely $x = 18$ and $y = 1$.

Now to return to the problem of finding d such that

$$e \times d \equiv 1 \bmod m$$

where e and m are given. What this means is that $e \times d$ is some multiple of m plus 1. In other words

$$e \times d = k \times m + 1 \qquad \text{or rearranging}$$
$$e \times d - m \times k = 1$$

Now this is precisely like $ax + by = c$ where the coefficients a and b are e and $-m$, the unknowns x and y are d and k and the constant c is 1.

So the trick is to apply Euclid's algorithm to e and m (the minus sign doesn't matter for the moment). Note that there will be a solution because we chose e to be relatively prime to m so that the gcd of e and m is 1 and of course 1 is a factor of the right hand side c which is also 1.

We now try this with $e = 5$ and $m = 264$.

$264 = 52 \times 5 + 4$	line 1
$5 = 1 \times 4 + 1$	line 2
$4 = 4 \times 1 + 0$	remainder 0 so gcd is last divisor = 1

We now work back

$1 = 5 - 1 \times 4$	using line 2 of gcd calculation
$= 5 - 1 \times (264 - 52 \times 5)$	using line 1
$= 53 \times 5 - 264$	gathering the terms in 5

So finally

$53 \times 5 = 1 \times 264 + 1$ or

$53 \times 5 \equiv 1 \mod 264$

and so 53 is the answer. Clearly this process can be automated.

Another example might be helpful. Suppose we use the primes $p = 97$ and $q = 101$ for the RSA algorithm. Then $n = 9797$ and $m = 96 \times 100 = 9600$. Now 9601 is prime so we cannot find a pair e and d that satisfy $e \times d \equiv 1 \mod 9600$ by using a pair of factors of 9601. We can choose e relatively prime to 9600 quite easily. Being bored with 7 we choose 11.

We then apply Euclid's algorithm to $e = 11$ and $m = 9600$ thus

$9600 = 872 \times 11 + 8$	line 1
$11 = 1 \times 8 + 3$	line 2
$8 = 2 \times 3 + 2$	line 3
$3 = 1 \times 2 + 1$	line 4
$2 = 2 \times 1 + 0$	remainder 0 so gcd is last divisor = 1

This requires rather more iterations than in the previous example. Then we unwind as before

$1 = 3 - 1 \times 2$	from line 4
$= 3 - 1 \times (8 - 2 \times 3)$	from line 3
$= 3 \times 3 - 8$	gathering 3s
$= 3 \times (11 - 1 \times 8) - 8$	from line 2
$= 3 \times 11 - 4 \times 8$	gathering 8s
$= 3 \times 11 - 4 \times (9600 - 872 \times 11)$	from line 1
$= 3491 \times 11 - 4 \times 9600$	gathering 11s

So rearranging we have

$3491 \times 11 = 4 \times 9600 + 1 \equiv 1 \mod 9600$

and we conclude that if $e = 11$ then $d = 3491$.

Clearly this algorithm is tedious to do by hand but can be programmed very easily.

The value of d may seem rather large but remember that $e \times d \equiv 1 \mod 9600$ and moreover, since 9601 is prime, the values for e and d must be such that $e \times d = 9600n + 1$ where n is greater than 1.

Taking $n = 2$, we have $19201 = 7 \times 13 \times 211$ so a possible pair of values is 91 and 211. And taking $n = 3$, we have $28801 = 83 \times 347$ so another modest pair is 83 and 347. Taking $n = 4$ in fact gives 11 and 3491 as obtained above.

Euler's function

THE FINAL loose end is to consider why it is that using d as described above does actually reverse the encryption process.

First we need to introduce the function $\phi(n)$ defined by Euler

$\phi(n)$ is the number of positive integers less than n and relatively prime to n

Remember that relatively prime means have no common factors so that the gcd with n is 1. For example, if n is 10, then the integers relatively prime to 10 are

1, 3, 7, 9

So $\phi(10) = 4$. Note that 1 is included because $\gcd(1, 10) = 1$.

Euler proved that if m is a positive integer and a is relatively prime to m then

$$a^{\phi(m)} \equiv 1 \bmod m$$

So, taking $m = 10$ and one of the integers relatively prime to 10 such as 7 then the theorem says that $7^{\phi(10)} \equiv 1 \bmod 10$. Now $\phi(10)$ is 4 and $7^4 = 2401$ which has remainder 1 on division by 10 so it works. Similarly $3^4 = 81$ and $9^4 = 6561$.

Before considering Euler's theorem in detail, it is worth looking at some of the properties of the function $\phi(n)$. The values of $\phi(n)$ for all values of n up to 60 are given in the table below. A number of things should be immediately obvious from the table.

$\phi(n)$ is even for all $n > 2$

$\phi(n)$ equals $n-1$ if n is prime

n	φ(n)	n	φ(n)	n	φ(n)	n	φ(n)	n	φ(n)
1	1	13	12	25	20	37	36	49	42
2	1	14	6	26	12	38	18	50	20
3	2	15	8	27	18	39	24	51	32
4	2	16	8	28	12	40	16	52	24
5	4	17	16	29	28	41	40	53	52
6	2	18	6	30	8	42	12	54	18
7	6	19	18	31	30	43	42	55	40
8	4	20	8	32	16	44	20	56	24
9	6	21	12	33	20	45	24	57	36
10	4	22	10	34	16	46	22	58	28
11	10	23	22	35	24	47	46	59	58
12	4	24	8	36	12	48	16	60	16

Values of Euler's function φ(n) for n up to 60.

The second is straightforward – if n is prime then all numbers less than n are relatively prime to it and there are $n-1$ of them. QED. Moreover, putting this into the theorem and changing n to p it reduces to

$$a^{p-1} \equiv 1 \bmod p$$

which is simply Fermat's Little Theorem which we met in the lecture on Notations. We will come back to the first observation later.

We recall from the lecture on Amicable Numbers that $\sigma(n)$ which is the sum of all the factors of n (including itself) has the property that

$$\sigma(m \times n) = \sigma(m) \times \sigma(n) \qquad \text{if } m \text{ and } n \text{ are relatively prime}$$

Euler's function $\phi(n)$ has exactly the same property

$$\phi(m \times n) = \phi(m) \times \phi(n) \qquad \text{if } m \text{ and } n \text{ are relatively prime}$$

Functions having this property are known as multiplicative functions. We can try it on a couple of examples. Thus 3 and 5 are relatively prime so it should apply to 3, 5, and 15; we have $\phi(3) = 2$, $\phi(5) = 4$, and $\phi(15) = 8$; and $8 = 2 \times 4$, so it does work. Consider 4 and 9, these are relatively prime although neither is prime itself; we have $\phi(4) = 2$, $\phi(9) = 6$, $\phi(36) = 12$, and again $12 = 2 \times 6$.

Another interesting property is that if n is the power of a prime, p^k, then

$$\phi(p^k) = p^k - p^{k-1} = p^k(1 - 1/p)$$

For example, consider $9 = 3^2$; we have $9 - 3 = 6$ which is correct. Similarly for $32 = 2^5$, we have $32 - 16 = 16$ which is correct. Note that the rule also works if $k = 1$, that is when n is actually a prime, since $p^{k-1} = p^0 = 1$, leading to $p-1$.

Armed with these two rules, we can now obtain $\phi(n)$ for any value of n. For example, for $n = 88$, we have $88 = 11 \times 2^3$. So

$$\phi(88) = \phi(11) \times \phi(2^3) = (11-1) \times (8-4) = 10 \times 4 = 40$$

Incidentally, it is now clear why $\phi(n)$ is even for all $n > 2$. For any prime $p > 2$, $\phi(p)$ is $p-1$ which is even. And for the power of any prime (including 2), it is $p^k - p^{k-1}$ and if p is 2 then both terms are even so the difference is even, whereas for any other prime both terms are odd so the difference is again even. So since $\phi(n)$ for any n is obtained by multiplying the ϕs for its factors, the result must be even.

We will now return to Euler's theorem that if m is a positive integer and a is relatively prime to m then

$$a^{\phi(m)} \equiv 1 \bmod m$$

Rather than actually formally prove it, we will illustrate the lines of the proof by an example. Suppose m is 10. Then the integers relatively prime to m are

1, 3, 7, 9

and we note that there are four of them since $\phi(10) = 4$. Take any one of these and multiply all of them by that number. Take 7 for example, we get

7, 21, 49, 63

now take the modulus of these by 10, that is the remainder on dividing by 10; we get

7, 1, 9, 3

and Lo and Behold that is the same set of numbers 1, 3, 7, 9 but in a different order.

It is not hard to see why. Since all the numbers are relatively prime to 10, the product of any two must be as well, so doing mod 10 must produce a number of the set. Moreover, no two can be the same since if they were, the difference between the originals would have to be a multiple of 10 which is impossible.

Now multiply all the numbers together, both the original set and the set after multiplication by 7. Clearly they are going to be the same mod 10. That is

$(1 \times 7) \times (3 \times 7) \times (7 \times 7) \times (9 \times 7) \equiv 1 \times 3 \times 7 \times 9 \bmod 10$

So gathering the 7s together we have

$7^4 \times 1 \times 3 \times 7 \times 9 \equiv 1 \times 3 \times 7 \times 9 \bmod 10$ or

$7^4 \times 189 \equiv 1 \times 189 \bmod 10$ where $189 = 1 \times 3 \times 7 \times 9$

Now remember the general rule that if a and m are relatively prime and if ab and ac are congruent mod m then we can cancel the a and deduce that b and c are congruent mod m as well. So in this case we conclude

$7^4 \equiv 1 \bmod 10$ since $\gcd(189, 10) = 1$

or in other words since $\phi(10) = 4$

$7^{\phi(10)} \equiv 1 \bmod 10$

and that's it.

Encryption and decryption

WE ARE NOW in a position to return to the RSA algorithm and see why d does the decryption corresponding to encryption with e. Remember that we have the encryption

$$c = v^e \bmod n$$

and then the decryption

$$w = c^d \bmod n$$

where we have written w for the moment. We need to show that w is always equal to v. Remember that we also have

$n = p \times q$ p and q are prime

$m = (p-1) \times (q-1)$

$e \times d \equiv 1 \bmod m$ e and d are "inverses" modulo m

Now

$$w = c^d \bmod n = (v^e)^d \bmod n = v^{e \times d} \bmod n$$

Moreover

$\phi(n) = \phi(p) \times \phi(q) = (p-1) \times (q-1) = m$, and so by Euler's theorem

$v^{\phi(n)} = v^m \equiv 1 \bmod n$ provided v is relatively prime to n

Now the fact that $e \times d$ is congruent to 1 mod m means that for some k, $e \times d = km+1$. So we can write w as

$$w = v^{e \times d} \bmod n = v^{km+1} \bmod n = (v^m)^k \times v \bmod n$$

Now (assuming that v is relatively prime to n), we just showed using Euler's theorem that v^m is congruent to 1 mod n and so we can remove the term $(v^m)^k$ and are left with

$$w = v \bmod n$$

or in other words since we are doing everything mod n it follows that $w = v$ and so we have shown that the decoding process works.

Note that we made the assumption that v was relatively prime to n. That will always be true unless v happens to be a multiple of one of the primes p or q. Even in that case it turns out that the algorithm still works as will now be shown.

Clearly the whole thing loses uniqueness if v exceeds n. But it is realistically possible that v might be a multiple of p or q but not both. Without loss of generality we will assume that v is a multiple of p but not of q.

So

$$v \equiv 0 \bmod p$$

and hence $v^{e \times d}$ is congruent to zero mod p as well. So we trivially have

$$w \equiv v^{e \times d} \equiv v \bmod p$$

Now since v is not a multiple of q it means that v is relatively prime to q. Hence we can use Fermat's Little Theorem that

$$v^{q-1} \equiv 1 \bmod q$$

Now remember that for some k, $e \times d = km+1 = k(p-1)(q-1)+1$; so

$$w = v^{e \times d} = v^{k(p-1)(q-1)+1} = v.(v^{q-1})^{k(p-1)} \equiv v \bmod q$$

So we have shown that w is congruent to v modulo both p and q. It immediately follows that w is congruent to v modulo $p \times q = n$. In other words

$$w \equiv v \bmod n$$

To see this remember what modulo means: $a \equiv b \bmod c$ means that c is some multiple of $b-a$. So if we also have $a \equiv b \bmod d$, then d must also be a multiple of $b-a$ and so it follows that cd must be a multiple of $b-a$ and so finally $a \equiv b \bmod cd$. Thus we have shown that the decryption works in all cases.

Code blocks

WE HAVE SEEN how the RSA algorithm can encrypt and decrypt numbers. The process will be unique provided the numbers are less than n.

In practice we want to encrypt text messages instead of raw numbers. We can do this by grouping several text items together in some way. Suppose we have code values of 1 for A, 2 for B and so on (we can use 0 for a space). Then we could encode a sequence of several letters such as CAT using base 27. Thus CAT would be

$$3 \times 27^2 + 1 \times 27^1 + 20 \times 27^0 = 3 \times 729 + 1 \times 27 + 20 = 2234$$

We can now encrypt $v = 2234$ and so on.

Suppose we use $p = 101$, $q = 197$ so that $n = 19897$ and $m = 19600$. Now $19601 = 17 \times 1153$ so we can chose $e = 17$ and $d = 1153$.

Using these values $v = 2234$ encrypts to $c = 10127$. This number could be sent as the encrypted message. Alternatively, we could turn this back into a string which might be more interesting. We do this by reversing the encoding process using base 27. We find

$$13 \times 27^2 + 24 \times 27 + 2 = 13 \times 729 + 24 \times 27 + 2 = 10127$$

and so the corresponding alphabetic sequence is MXB.

Thus a long message such as "SEND MORE MONEY" can be split into groups of three letters thus SEN D_M ORE _MO NEY where we have represented the spaces by an underscore. The 5 groups are encoded to the 5 numbers 14000, 2929, 11426, 366, 10366. These are then encrypted to give the 5 numbers 5495, 9090, 4925, 15807, 1721 and these can be used as the message or we can decode to give GNN, LLR, FTK, URL, BIT.

A minor point to notice with this is that 27^3 is 19683 which is slightly less than n = 19897. So if we always group letters into threes, then some possible values will not be used. This is not a problem but note that the encrypted message could be larger than 19682 such as 19777 and so if we converted that back into an alphabetic group we would get a four letter group, in this case A_CM.

If we wanted to use the ten digits as well as letters then we could use base 37. We could also use the full 7-bit or 8-bit ISO codes but we could not realistically turn any numeric sequence back into a letter sequence such as GNN, LLR etc. because the decoding process would give rise to characters with Latin-1 values which do not correspond to a printable character. For example, the value 7 is meant to ring the bell on equipment such as a traditional telex machine which has a bell.

We conclude this discussion on the RSA algorithm by pointing out that in practice, p and q are chosen to be very large, perhaps with 100 digits each. It is fairly easy to find such large primes but quite impossible to factorize the resulting 200 digit value for n in a reasonable time with current technology. If computers get very much faster we can just choose larger numbers for p and q.

Animal gaits

AND NOW for something quite different. The gaits of animals, especially quadrupeds, illustrate a number of possible symmetries. But we start by considering bipeds such as us.

Typically we move by first moving one leg and then the other in a symmetric manner. We could denote this by something like

LR LR LR ... or 12 12 12 ...

which echo the Left Right Left Right or One Two One Two commands of the old drill sergeant.

Walking and running both use the same pattern. The difference lies in the number of feet in contact with the ground. In walking there are 1 or 2 feet on the ground whereas in running there are 0 or 1 feet on the ground at any time.

The other possible symmetric motion for bipeds is jumping with both feet together. This might be denoted by

(LR) (LR) ... or (12) (12) ...

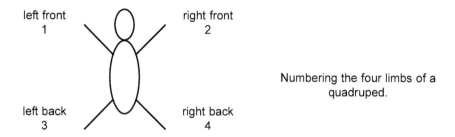

Numbering the four limbs of a quadruped.

where we bracket together the two legs which move together. Such jumping is very inefficient for humans but kangaroos do it.

Small birds such as sparrows and dunnocks jump but larger birds such as magpies and blackbirds walk.

Many possibilities arise with four legs. We denote the legs by 1234 where 1 and 2 are the front legs and 1 and 3 are the left legs as in the diagram above.

First we can extend the walking/running and jumping motion of bipeds to quadrupeds with each back leg moving in the same way as the corresponding front leg. Thus we get

$$(13)(24)\ (13)(24)\ \ldots \qquad \text{pace}$$

This is known as the pace. Not many quadrupeds do this. But camels and giraffes do it. Early postage stamps of the Sudan show a mail camel pacing along. The lecture on Notations shows examples of such stamps.

Another extension of the biped walk is where both back legs are together and both front legs are together. This gives

$$(12)(34)\ (12)(34)\ \ldots \qquad \text{bound}$$

which is known as the bound. Squirrels do this. And so do horses in the imagination of many historic painters.

We can also extend the biped jump in a similar way. Thus

$$(1234)\ (1234)\ \ldots \qquad \text{pronk}$$

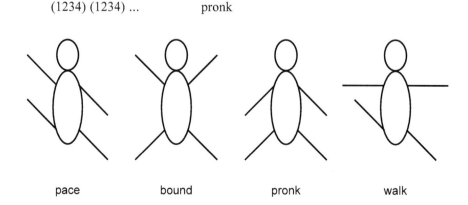

pace   bound   pronk   walk

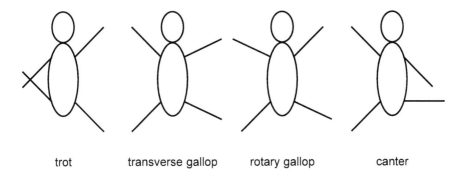

trot   transverse gallop   rotary gallop   canter

This four-legged jump is known as the pronk. It is said that deer do it when startled.

Now consider the extra possibilities introduced by having four legs. For example

    3142 3142 ...   walk

    (23)(14) (23)(14) ...   trot

These are both done by horses. In the case of the walk, only one leg is ever off the ground at a time, and sometimes all four legs are on the ground. It is the steady gait of old dobbin pulling a heavy cart.

The trot is much more lively and the hooves make a jolly rattle on the cobbles. The diagonal pairs of legs work together in a symmetric manner. Sometimes all four legs are on the ground and sometimes only two.

There are two possible forms of gallop

    3412 3412 ...   transverse gallop

    3421 3421 ...   rotary gallop

Note that, as in the walk, no two legs are in synchrony; nevertheless the rhythm is quite different. In the walk the four legs move with a uniform time interval between them whereas in the gallops, the back legs are close together in time as are the front legs. The times are 0, 0.1, 0.5, 0.6, 1.0, 1.1, ... The transverse gallop is done by horses and dogs. The rotary gallop is done by a camel in a hurry. Sometimes there are no legs on the ground and sometimes just one.

Finally, there is the canter which seems a bit odd

    14(23) 14(23) ...   canter

Note that the front right and back left leg go together but the front left and back right do not. The number of legs on the ground at any time is probably either one or two.

There are clearly opposite variants of both gallops and canter for left-handed horses and camels which are

4321 4321 ...	opposite transverse gallop
4312 4312 ...	opposite rotary gallop
23(14) 23(14) ...	opposite canter

The canter is done by horses but not by camels.

Horses thus have four modes, the walk, the trot, the canter, and the (transverse) gallop. We can think of horses as having four forward gears in ascending order of speed. Camels seem to have just three forward gears, namely, the walk, the pace, and the (rotary) gallop.

Which gait is used by an animal at any time seems to be that which uses least effort for the speed concerned.

Towers and rings

WE CONCLUDE with two (hopefully) amusing topics. One is the Tower of Hanoi and the other is known as Chinese Rings. They are both very old and have interesting similarities and differences.

The Tower of Hanoi is sometimes phrased in terms of monks in a temple moving a tower of 64 golden discs between two diamond needles using a third needle as an intermediary. The discs are of different sizes and a disc may not be placed on top of a smaller one; they are moved one at a time. It is said that when they finish moving the discs the world will end with a mighty thunderclap.

Let us attempt a smaller example with perhaps just four wooden discs on three steel pins. We will call the pins A, B, and C and number the discs 1, 2, 3, and 4 with 1 being the smallest. Suppose we start with the tower of discs on pin A and wish to move it to pin C.

This is an example of a recursive algorithm. To move the four discs, first move the top three (123) to the intermediate pin B. Then move the bottom disc 4 to its final place on pin C, then move the three discs from B to C. To move the three discs from A to B we use the same process – first move the top two (12) to pin C, then move disc 3 to pin B and then move (12) from C to B. And of course to move the two discs (12) from A to C we just move 1 from A to B, then 2 from A to C and then 2 from C to B.

The diagram opposite shows the details of moving the three discs (123) from A to B and moving the bottom disc 4 to its final resting place on C. Moving the three discs from B to C follows a similar pattern.

Suppose that the total number of moves required to move n discs is $F(n)$. Then it is easy to show that $F(n) = 2^n - 1$.

10 Finale   221

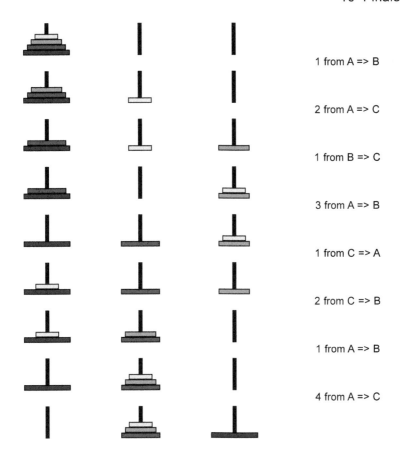

The Tower of Hanoi.

To move n discs we have to move $n-1$ discs twice and then disc n. In other words $F(n) = 2F(n-1) + 1$. We note that

$$2 \times (2^{n-1} - 1) + 1 = 2 \times 2^{n-1} - 2 + 1 = 2^n - 1$$

which proves it. So for 4 discs the number of moves is 15.

This puzzle is very easy to program but it is also easy to make a mistake doing it by hand. From a general position there are three possible moves – disc 1 can be moved to either of the other pins and a disc can be moved between the two pins that disc 1 is not on. Of these three moves, one is the reverse of the previous move and would take us back a step, and of the other two, one is correct and the other is wrong. Another point to note is that alternate (correct) moves involve disc 1.

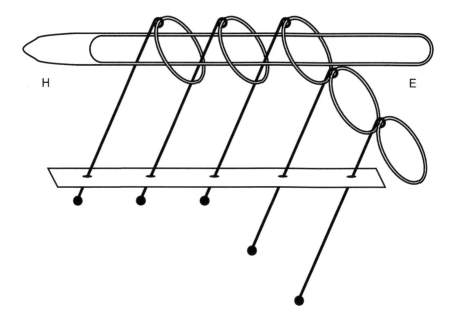

The Chinese rings.

It is interesting to compare the Tower of Hanoi with the puzzle often known as Chinese Rings which is very old. The figure above shows the general arrangement with five rings. A typical puzzle might have as many as twelve rings.

It consists of a looped bar on a handle H. On the bar are a number of rings and each ring has a loop of wire attached to it which goes through the adjacent ring and finally through a strip of board.

The arrangement is such that the ring at the end E can be taken on and off the bar at any time. However, any other ring can only be moved on or off if the one next to it towards the end E is on the bar and all the others towards E are off the bar. The order of the rings obviously cannot be changed.

If the bar has n rings then it takes about $2^{n+1}/3$ moves to take all the rings off. If we can make one move per second and the bar has 12 rings then it takes about 45 minutes to take the rings off and another 45 minutes to put them on again. That is provided no mistakes are made!

Assume that the rings are numbered from the right so that the end ring is number 1. It is actually possible to move rings 1 and 2 together but for the sake of the discussion we shall assume that this is not done and that it is only possible to move ring 2 off or on if ring 1 is on. So in numerical terms the rule is that ring n can only be moved if ring $n-1$ is on and rings 1 to $n-2$ are all off.

An important point is that at any position there are always just two moves that can be made. For example, in the position shown ring 1 can be put on and ring 4 can be taken off. Nothing else can be done. Note that ring 1 can always be moved and so only one other ring can ever be moved at any time.

If we start with all the rings on then our first move can only be either to take ring 1 off or to take ring 2 off. If we have 5 rings as above then in order to get ring 5 off we must have ring 4 on and rings 1, 2, and 3 off. It is clear therefore that our first subgoal must be to get ring 3 off. In order to do this we need ring 2 on and ring 1 off.

So our first move is to take ring 1 off. Then we can take ring 3 off. Now we need to get ring 2 off and in order to do that we put ring 1 on, then take ring 2 off and then we take ring 1 off again. Now we have rings 1, 2, and 3 off and ring 4 is on so we can now take off ring 5. The only ring on now is 4 and in order to get that off we need ring 3 on and rings 2 and 1 off. And so on. If we depict the position as a sequence of 1 and 0 with 1 denoting a ring on then the sequence is

11111	all on at start
11110	
11010	
11011	
11001	*
11000	
01000	take 5 off
01001	*
01011	
01010	
01110	
01111	
01101	*
01100	
00100	take 4 off
00101	*
00111	
00110	
00010	take 3 off
00011	
00001	* take 2 off
00000	and done

That is 21 moves. Because there only two possible moves at any time and one must take us back to where we were before, there is in fact only one possible forward move that can be taken. And so provided we do not accidentally go into reverse, it is quite straightforward.

If we allow ourselves to move rings 1 and 2 together then those 5 moves marked with an asterisk can be combined with the next move giving just 16 moves for 5 rings.

We will now find a general formula $F(n)$ for the number of steps required to take n rings off. First we take off $n-2$ rings (that takes $F(n-2)$ steps). Then take off ring n (1 step). Now put back rings 1 to $n-2$ (that takes $F(n-2)$ steps). All the rings are now on except for ring n. So to remove these will take $F(n-1)$ steps. The sequence of play is

1111…11	n rings on at start	
1100…00	remove $n-2$ rings	$F(n-2)$ steps
0100…00	remove ring n	1 step
0111…11	put back $n-2$ rings	$F(n-2)$ steps
0000…00	remove $n-1$ rings	$F(n-1)$ steps

These individual moves must add up to $F(n)$. So we have

$$F(n) = F(n-2) + 1 + F(n-2) + F(n-1) = F(n-1) + 2F(n-2) + 1$$

This is a difference equation. We can solve it by using the technique explained in Appendix G. This gives the number of moves required as

$F(n) = (2^{n+1} - 1)/3$ n odd

$F(n) = (2^{n+1} - 2)/3$ n even

We can check that for $n = 5$ and get $F(5) = (2^6 - 1)/3 = 63/3 = 21$ which agrees with the analysis above.

Note that if we allow ourselves to move rings 1 and 2 together, then the number of moves becomes

$F(n) = 2^{n-1}$ n odd

$F(n) = 2^{n-1} - 1$ n even

which for $n = 5$ becomes 16 which again agrees with the analysis above.

So like the Tower of Hanoi, the number of moves required doubles every time an extra item (disc or ring) is added. But unlike the Tower of Hanoi, at any position in the case of the rings only two moves are possible – one back to our previous position and one forward. So provided we do not go backwards by mistake it is relatively easy to do even though it looks gruesome.

Further reading

FOR A GENERAL DESCRIPTION of the background to the RSA algorithm see *The Code Book* by Simon Singh although the description of the algorithm itself is rather thin. For a more mathematical description see *Elementary Number Theory and its Applications* by Kenneth Rosen.

A brief description of animal gaits will be found in *Nature's Numbers* by Ian Stewart. For a fuller description of the (hard) mathematics behind this topic see for example *Nonlinear dynamics of networks: the groupoid formalism* by Martin Golubitsky and Ian Stewart in Bulletin of the American Mathematical Society, July 2006.

Chinese rings are explained in *Mathematical Recreations and Essays* by Rouse Ball and Coxeter. They are also discussed in *Amusements in Mathematics* by Dudeney.

See also Appendix G on Differences for a more detailed discussion on the solutions to the Towers and Rings and other similar puzzles and Appendix H on the Chinese Remainder Theorem which has a further discussion on the Diophantine equations involved in the RSA algorithm.

A Ackermann

WHEN DISCUSSING favourite numbers we mentioned the factorial function and the Ackermann function. The factorial function is well known. This appendix explains the very curious Ackermann function.

Recursion and iteration

WE CAN DEFINE the factorial function by the following mathematical statements which define it in terms of itself

$$\text{factorial}(n) = n \times \text{factorial}(n-1), \quad n > 0,$$
$$\text{factorial}(0) = 1$$

We also informally write

$$n! = n \times (n-1) \times (n-2) \times \cdots \times 3 \times 2 \times 1$$

We can program this on a computer in one of two ways. We can follow the formal mathematical definition and write something like

```
function Factorial(n: Integer) return Integer is
begin
  if n = 0 then
    return 1;
  else
    return n × Factorial(n–1);
  end if;
end;
```

This is known as a recursive definition because the function uses itself. The alternative way is to follow the informal approach and write

```
function Factorial(n: Integer) return Integer is
  Answer: Integer := 1;
begin
  for k in 1 .. n loop
    Answer := Answer × k;
  end loop;
  return Answer;
end;
```

in which we multiply all the integers together starting from 1. This is an iterative definition. Using a recursive definition is usually considered more elegant but an iterative approach is often used because it is felt to be quicker.

The mathematical function known as Ackermann is interesting. It was devised by the German mathematician Wilhelm Ackermann (1896–1962) and colleagues as part of research on computability. There were various versions but the one usually referred to as Ackermann is defined as follows

$A(0, n) = n+1$,

$A(m, 0) = A(m-1, 1)$, $\quad m > 0$,

$A(m, n) = A(m-1, A(m, n-1))$, $\quad m > 0$ and $n > 0$

We can program this quite easily recursively

```
function A(m, n: Integer) return Integer is
begin
  if m = 0 then
    return n+1;
  elsif n = 0 then
    return A(m-1, 1);
  else
    return A(m-1, A(m, n-1));
  end if;
end;
```

The curious feature of the Ackermann function is that we cannot convert the recursion into loops (that is, unroll) as we can with the factorial function. In other words it cannot be done in an iterative fashion.

The values of Ackermann for quite modest values of m and n can be huge. Thus $A(3, 6)$, one of the favourite numbers mentioned in the first lecture, equals 509. But $A(4, 2)$ is far greater than the number of atoms in the universe.

m	0	1	2	3	4	5	6	n
0	1	2	3	4	5	6	7	$n + 1$
1	2	3	4	5	6	7	8	$2+(n+3) - 3$
2	3	5	7	9	11	13	15	$2\times(n+3) - 3$
3	5	13	29	61	125	253	509	$2^{(n+3)} - 3$
4	13	65533	$2^{65536}-3$	...	...	...	...	$2^{2^{2^{\cdot^{\cdot^{\cdot^{2}}}}}} - 3$

Some values of Ackermann's function.

An intriguing feature is that the recursive definition of Ackermann only involves addition and subtraction and recursion. Factorial gets large because of the multiplication. But Ackermann gets large purely because of the recursion.

The table opposite shows the values of the function for some small values of m and n. Note in particular the general form of A(m, n) for small values of m. They involve 2 operated upon in some way by (n+3) followed by the subtraction of 3. When n is 1, the operation is addition, when n is 2, the operation is multiplication, when n is 3, the operation is exponentiation, and when n is 4, it becomes a tower of exponents. Thus 13 is $2^{2^2} - 3$ and so on.

It important to note that the function always terminates. In each step, either m decreases, or m is unchanged and n decreases. Whenever n becomes zero, m is decreased and so m eventually becomes zero as well. However, when m is decreased, the new initial setting for n can be very large indeed and so the whole thing takes a long time.

The values of Ackermann get very large but the number of internal recursive calls, which clearly dominates the time taken, becomes even larger. The table below shows the number of calls for various values of m and n. Thus in order to compute the favourite number A(3, 6) which is 509, the internal workings of the recursion cause A to be called 172,233 times!

The almost catastrophic way in which things get out of hand is illustrated by considering A(4, 1). From the definition we have

$$A(m, n) = A(m-1, A(m, n-1))$$

so putting $m = 4$ and $n = 1$ we get

$$A(4, 1) = A(3, A(4, 0)) = A(3, 13) = 2^{16} - 3 = 65533$$

From the table below, we note that the number of internal calls required in order to compute A(3, n) roughly increases by 4 for each increment to n. So to compute A(3, 13) there are over 2000 million calls and therefore just a few more to compute A(4, 1); to be precise 2,862,984,010.

m	0	1	2	3	4	5	6	n
0	1	1	1	1	1	1	1	1
1	2	4	6	8	10	12	14	$2n+2$
2	5	14	27	44	65	90	119	$2n^2+7n+5$
3	15	106	541	2432	10307	42438	172233	?
4	107	2862984010	...	...	...	...	...	???

Number of calls of A required to compute the final value.

m	0	1	2	3	4	5	6	n
0	1	1	1	1	1	1	1	1
1	2	3	4	5	6	7	8	$n+2$
2	4	6	8	10	12	14	16	$2n+4$
3	7	15	31	63	127	255	511	$2^{n+3}-1$
4	16	65536	2^{65536}	...	...	...	...	$2^{2^{2^{\cdots^{2}}}}$

Maximum nested calls of A required to compute the final value.

There are two possible difficulties with computing A on a typical small computer. One is that we might generate a number which is too large for the implementation in which case we will get an overflow condition. The other difficulty is that we might run out of space. This is because not only do lots of calls of A occur but many will be pending at any time awaiting for others to complete. It is easiest to see this with factorial. If we call **Factorial(8)** then the first thing that happens is that the code promptly calls **Factorial(7)**; this means that the state of the first call has to be preserved somewhere while the second call is processed. In fact all the calls are in suspense until the code calls **Factorial(0)**; this causes them all to unwind and do all the multiplications. The storage for preserving these states can run out and although in the case of factorial, overflow will undoubtedly occur first, in the case of Ackermann, storage is likely to run out first.

The maximum nested depth for Ackermann is shown above and it will be noted that it differs from the final value by just $m-1$. On a typical small computer, $A(4, 1)$ fails because it runs out of space.

The Ackermann function has been used to test compilers. A compiler is a program that converts a program in textual form into the binary pattern that is actually executed by the computer. One of the first examples of such testing was using the Whetstone Algol 60 compiler on the KDF9 computer in the 1960s. The KDF9 was one of the first solid state computers and was passionately adored by many programmers.

Further reading

THE WIKIPEDIA entry for Ackermann contains a good description. See http://en.wikipedia.org/wiki/Ackermann_function. There is an interesting description of related giant numbers in *The Book of Numbers* by Conway and Guy.

B Pascal's Triangle

WE BRIEFLY met Pascal's triangle in the lecture on Probability when we saw how it arose through the ways of getting different numbers of heads and tails when tossing several coins. We also noted that it could be introduced as the coefficients in the binomial expansion of $(x + y)^n$.

Pascal's triangle has many intriguing properties and is worth exploring in some detail.

Basic properties

FIRST WE RECALL the basic properties that each number is the sum of the two in the row above and that the numbers in the rows add up to the powers of 2 as shown below.

The sequences of numbers in the diagonals are interesting and are easily deduced from the fact that each number is the sum of the two above.

The first diagonal is all 1s. The second diagonal has the integers in order. This is simply because each is formed by adding 1 to its predecessor in the row above.

```
                            1                               1
                          1   1                             2
                        1   2   1                           4
                      1   3   3   1                         8
                    1   4   6   4   1                      16
                  1   5  10  10   5   1                    32
                1   6  15  20  15   6   1                  64
              1   7  21  35  35  21   7   1               128
            1   8  28  56  70  56  28   8   1             256
          1   9  36  84 126 126  84  36   9   1           512
        1  10  45 120 210 252 210 120  45  10   1        1024
      1  11  55 165 330 462 462 330 165  55  11   1      2048
    1  12  66 220 495 792 924 792 495 220  66  12   1    4096
```

The rows of Pascal's triangle add to the powers of 2.

232 Nice Numbers

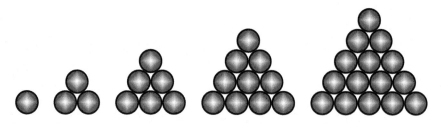

The triangular numbers.

The third diagonal has the sequence

1, 3, 6, 10, 15, 21, 28, 36, 45, ...

These are the so-called triangular numbers because they are the number of balls (cannonballs perhaps) that can be arranged in triangles of various sizes as shown above. The triangular number of size n is clearly n more than the that of size $n-1$. We easily deduce that the triangular numbers are given by the formula

$T_n = n(n+1)/2$

There is an amusing story about the German mathematician Carl Gauss (1777–1855) as a young boy at school. The master is said to have set the class the task of adding the numbers from 1 to 100 together hoping that would keep them quiet for a bit. However, young Gauss very quickly calculated the answer in his head much to the astonishment of the master. What he did mentally was write them in order and then on the line below write them out backwards and then add the two rows together thus

$$\begin{array}{rl} T = & 1 + 2 + 3 + 4 + \cdots + 97 + 98 + 99 + 100 \\ T = & 100 + 99 + 98 + 97 + \cdots + 4 + 3 + 2 + 1 \\ 2T = & 101 + 101 + 101 + 101 + \cdots + 101 + 101 + 101 + 101 \\ = & 101 \times 100 = 10100 \end{array}$$

and so finally $T = 5050$. The general formula for T_n follows by using the same technique.

The next diagonal of Pascal's triangle has the numbers

1, 4, 10, 20, 35, 56, 84, 120, ...

and these are tetrahedral numbers because they are the number of balls that can be arranged in tetrahedra of various sizes as shown opposite. Each tetrahedral number is the sum of all the triangular numbers up to that. We can also refer to them as triangular pyramidal numbers. The general formula is

$TP_n = n(n+1)(n+2)/6$

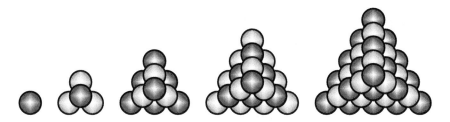

The tetrahedral numbers, or triangular pyramidal numbers.

And so it goes on – the next diagonal is the series of numbers representing the sizes of 4-simplexes (the analogues of tetrahedra) in four dimensions. Thus

1, 5, 15, 35, 70, 126, 210, ...

Each is the sum of the tetrahedral numbers up to it just as each tetrahedral number is the sum of the triangular numbers. The general formula is

$TQ_n = n(n+1)(n+2)(n+3)/24$

Looking back at the lecture on Probability, we see that we are just rediscovering the formula for the number of ways of choosing r things from n things which was

$$^nC_r = \frac{n!}{(n-r)! \times r!} \quad \text{where } n! = n \times (n-1) \times (n-2) \times (n-3) \times \cdots \times 2 \times 1$$

The triangular numbers correspond to $r = 2$, the tetrahedral numbers correspond to $r = 3$, and the fourth dimensional numbers correspond to $r = 4$.

One can find other apparently curious properties of the Pascal diagonals. For example consider that starting 1, 3

1, 3, 6, 10, 15, 21, 28, ...

they all divide by 3 except the fourth and seventh. In the case of the 1, 5 sequence which starts

1, 5, 15, 35, 70, 126, 210, ...

they all divide by 5 except the sixth (and eleventh) and in the case of the 1, 7 sequence which starts

1, 7, 28, 84, 210, 462, 924, 1716, ...

they all divide by 7 except the eighth (and fifteenth). This might at first sight seem to be some mysterious property of prime numbers but it is all a consequence of the general formula for ^nC_r given above.

Fibonacci numbers

REMEMBER the sequence of Fibonacci numbers where each is the sum of the previous two? These were introduced in the lecture on Amicable Numbers where we saw that the ratio of pairs of Fibonacci numbers converges to the golden number. The sequence starts

1, 1, 2, 3, 5, 8, 15, 21, 34, 55, 89, ...

We find that these numbers are also lurking in Pascal's triangle as shown below. The slightly offset diagonals have sequences such as 1, 5, 6, 1 and 1, 7, 15, 10, 1 (in blue) and these sum to the alternate Fibonacci numbers 13 and 34 whereas between these lie sequences such as 1, 4, 3 and 1, 6, 10, 4 (in red) which sum to the other Fibonacci numbers such as 8 and 21.

The reason is not hard to see. Consider the second red band which is 1, 6, 10, 4. The 6 is 1 plus 5 in the line above – the 1 is in the upper red band and the 5 is in the blue band. Similarly the 10 is 4 plus 6 where the 4 is in the upper red and the 6 in the blue band; and the 4 is 3 plus 1 where the 3 is in the upper red and the 1 in the blue band. That leaves just the 1 in the second red band which matches the 1 in the blue band. So everything in the second red band is matched by adding all numbers in the two bands above. That is just the rule for Fibonacci numbers; the only other thing to check is that the sequence starts correctly which it does.

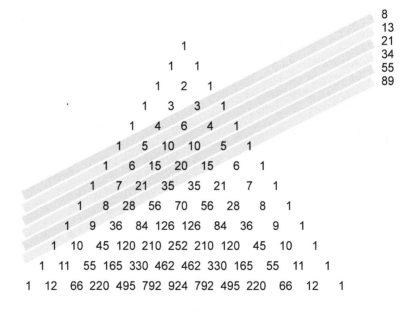

The offset diagonals sum to the Fibonacci numbers.

Squares and pyramids

IT IS CLEAR that square and higher numbers can be visualized in much the same way as the triangular numbers. The first thing to note is that the sum of consecutive triangular numbers is a square number as illustrated below. Thus we have 1+3 = 4, 3+6 = 9, 6+10 = 16, 10+15 = 25, and so on.

More generally, we saw above that the triangular numbers are given by

$T_n = n(n+1)/2$

and so adding the values of T_n and T_{n-1} we just get

$T_n + T_{n-1} = n(n+1)/2 + (n-1)n/2 = (n^2 + n + n^2 - n)/2 = n^2 = S_n$

Square pyramidal numbers can be made by adding together square numbers as shown below, just as triangular pyramidal numbers (tetrahedral numbers) can be formed from adding together triangular numbers as shown earlier.

Moreover, since each square layer comprises the sum of two triangles, it follows that a square pyramid must also be the result of adding together adjacent triangular pyramids. This is illustrated overleaf. We saw earlier that the triangular pyramidal numbers which we denote by TP_n are given by

$TP_n = n(n+1)(n+2)/6$

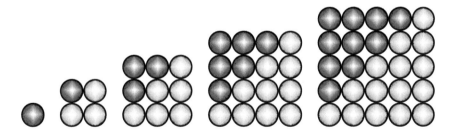

Consecutive triangular numbers make a square number.

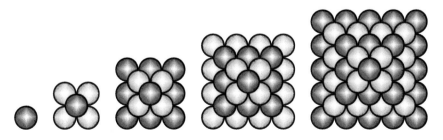

The square pyramidal numbers.

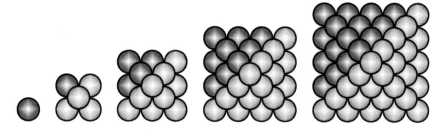

Consecutive triangular pyramids make a square pyramid.

So the square pyramidal numbers SP_n must be given by

$$SP_n = TP_n + TP_{n-1} = n(n+1)(n+2)/6 + (n-1)n(n+1)/6 \quad \text{giving}$$
$$SP_n = n(n+1)(2n+1)/6$$

So let us summarize the relationships we have seen. They are

$$T_n = T_{n-1} + n \qquad S_n = T_n + T_{n-1}$$
$$TP_n = TP_{n-1} + T_n \qquad SP_n = SP_{n-1} + S_n$$
$$SP_n = TP_n + TP_{n-1}$$

The table below gives the first few values.

Puzzle 138 from Dudeney concerns cannonballs. Soldiers are ordered to arrange their cannonballs into square pyramids. Moreover, each pyramid is to have a square number of balls. Neglecting the trivial solution of a pyramid of just one ball, this means we are asked to find a number which is both a square number and a square pyramidal number. The answer is 4900 which is the 24th square pyramidal number as well as being 70 squared. Using the formula above for SP_n and setting $n = 24$ we have

$$SP_{24} = 24 \times 25 \times 49 / 6 = 4 \times 25 \times 49 = 4900 = 70^2$$

The reader is invited to find another such number which is both a square pyramidal number and a square number (or to show that no other such number exists).

Natural n	1	2	3	4	5	6	7	8	9
Triangular T_n	1	3	6	10	15	21	28	36	45
Square S_n	1	4	9	16	25	36	49	64	81
Triang pyramid TP_n	1	4	10	20	35	56	84	120	165
Square pyramid SP_n	1	5	14	30	55	91	140	204	285

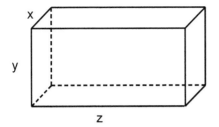

A rectangular box with sides x, y, z.

Another interesting square number puzzle concerns Pythagorean triples. We know that there are many ways of making a rectangle with integer sides and whose diagonal is also an integer. Thus sides of length 3 and 4 give rise to a diagonal of length 5 since $3^2 + 4^2 = 5^2$. Another example is sides 7 and 12 giving diagonal 13.

Now consider the equivalent problem in three dimensions. Is it possible to have a rectangular box whose sides are all integers and where the diagonals of the faces are also integers? In other words can we find integers, x, y, and z such that $x^2 + y^2$ and $x^2 + z^2$ and $y^2 + z^2$ are all perfect squares?

The answer is yes. The smallest such set of numbers is 44, 117, and 240. We have

$$44^2 + 117^2 = 125^2$$
$$44^2 + 240^2 = 244^2$$
$$117^2 + 240^2 = 267^2$$

However, the body diagonal is not an integer since $44^2 + 117^2 + 240^2 = 73225$ and this is not a perfect square. So this raises the question of whether it is possible for the body diagonal to be an integer as well. In other words for $x^2 + y^2 + z^2$ also to be a perfect square. This seems to be an unsolved problem. There are many examples of two face diagonals and the body diagonal being integers. The smallest is with sides 448, 495, and 840. We have

$$448^2 + 840^2 = 952^2$$
$$495^2 + 840^2 = 975^2$$
$$448^2 + 495^2 + 840^2 = 1073^2$$

but $448^2 + 495^2 = 445729$ which is not a perfect square so the other face diagonal is not an integer.

Further reading

SEE *Amusements in Mathematics* by Dudeney for the puzzle regarding the pile of cannonballs.

C Stochastics

WE ENCOUNTER many situations in everyday life where events occur in sequence and have a probabilistic flavour. A good example is waiting for a bus. Buses should arrive at regular intervals but for various reasons they often seem to arrive at random or perhaps in a bunch.

Other examples where time is not so involved occur in many gaming situations. In this appendix we look at examples of both these kinds which are known as stochastic processes.

War games

FIRST CONSIDER a simplified version of the game Risk. There are two territories. One is occupied by several armies of A, the Attacker. The other is occupied by several armies of D, the Defender. At each turn they both roll a die. That with the higher score wins and one army of the loser is removed. In the event of a draw the Defender wins.

The game continues until either the defender loses all armies in which case the attacker then moves one or more armies into the territory, or the attacker has only one army left. An important rule is that the attacker cannot fight with only one army since an army must be moved into the defended territory when it is conquered. The defender clearly has an advantage because of the rule that in the event of a draw, the defender wins. A typical question is, if the defender has five armies how many armies must the attacker have in order that the probability of winning is perhaps 60%?

From the table below it is clear that the attacker wins in 15 cases whereas the defender wins in 21 cases out of the total of 36 equally likely outcomes.

The probability of the attacker winning at each stage is therefore $5/12$. And the probability of the defender winning is $7/12$.

D\A	1	2	3	4	5	6
1	D	A	A	A	A	A
2	D	D	A	A	A	A
3	D	D	D	A	A	A
4	D	D	D	D	A	A
5	D	D	D	D	D	A
6	D	D	D	D	D	D

The attacker wins in the red cases and the defender in the blue cases.

240 Nice Numbers

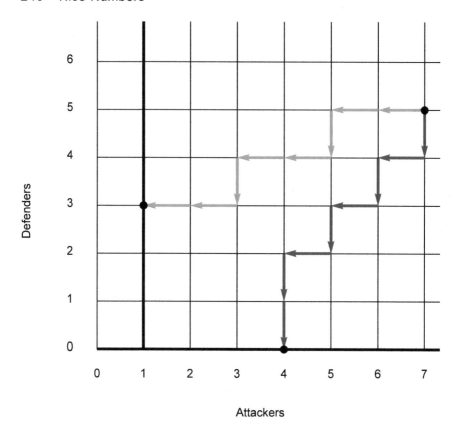

At each stage the situation can be represented by a pair of numbers such as (7, 5) which corresponds to the state where the attacker has 7 armies and the defender has 5 armies. Each time the dice are rolled one of them loses an army so the situation will move to the state (7, 4) (with probability $5/12$) or to the state (6, 5) (with probability $7/12$).

The progress of a typical game can be traced on the diagram above. Thus it might go (7, 5), (7, 4), (6, 4), (6, 3), (5, 3), (5, 2), (4, 2), (4, 1), (4, 0) as shown in red in which case the attacker wins. Or it might go (7, 5), (6, 5), (5, 5), (5, 4), (4, 4), (3, 4), (3, 3), (2, 3), (1, 3) as shown in blue in which case the defender wins. Remember that the attacker cannot carry on with only one army so the game stops when the attacker has one army or the defender has no armies – these barrier lines are shown in bold.

As we saw above, the probability of a move down is $5/12$ and the probability of a move to the left is $7/12$. Starting at (7, 5) the only possible moves are to (7, 4) with probability $5/12$ and to (6, 5) with probability $7/12$. It should be clear therefore that the probability of the attacker winning from (7, 5) must equal the probability of winning from (7, 4) multiplied by $5/12$ plus the probability of winning from (6, 5) multiplied by $7/12$.

C Stochastics

In general if we denote by $P_{m,n}$ the probability of the attacker winning from the position (m, n) then we have

$$P_{m,n} = \lambda P_{m,n-1} + \mu P_{m-1,n} \quad \text{where } \lambda = 5/12 \text{ and } \mu = 7/12$$

This equation is known as the Chapman–Kolmogorov equation after the British mathematician Sydney Chapman (1888–1970) and the Russian mathematician Andrei Kolmogorov (1903–1987).

Now clearly the probability of the attacker winning from the state (2, 1) is $5/12$ so we have

$$P_{2,1} = 5/12$$

Note also that there are terminal cases when $m = 1$ and $n = 0$

$P_{1,n} = 0$ attacker has only one army, cannot play

$P_{m,0} = 1$ defender has lost all armies, cannot play

We can then work out the higher values thus

$$P_{2,2} = \lambda P_{2,1} + \mu P_{1,2} = 5/12 \times 5/12 + 7/12 \times 0 = 25/144 = 0.1736...$$

$$P_{3,1} = \lambda P_{3,0} + \mu P_{2,1} = 5/12 \times 1 + 7/12 \times 5/12 = 95/144 = 0.6597...$$

$$P_{2,3} = \lambda P_{2,2} + \mu P_{1,3} = 5/12 \times 25/144 + 7/12 \times 0 = 125/1728 = 0.0723...$$

and so on. This is easily programmed and we get the results shown in the table below.

So if the defender has five armies then the attacker needs nine to have a 60% chance of success.

					Attackers					
Ds	1	2	3	4	5	6	7	8	9	10
1	0.0000	0.4167	0.6597	0.8015	0.8842	0.9325	0.9606	0.9770	0.9866	0.9922
2	0.0000	0.1736	0.3762	0.5534	0.6912	0.7917	0.8621	0.9100	0.9419	0.9629
3	0.0000	0.0723	0.1989	0.3466	0.4902	0.6158	0.7185	0.7983	0.8581	0.9018
4	0.0000	0.0301	0.1005	0.2030	0.3227	0.4448	0.5588	0.6586	0.7417	0.8084
5	0.0000	0.0126	0.0492	0.1133	0.2005	0.3023	0.4092	0.5131	0.6084	0.6917
6	0.0000	0.0052	0.0235	0.0609	0.1191	0.1954	0.2845	0.3798	0.4750	0.5653
7	0.0000	0.0022	0.0111	0.0319	0.0682	0.1212	0.1893	0.2686	0.3546	0.4424
8	0.0000	0.0009	0.0051	0.0163	0.0379	0.0726	0.1212	0.1826	0.2543	0.3327
9	0.0000	0.0004	0.0024	0.0082	0.0206	0.0423	0.0752	0.1199	0.1759	0.2412
10	0.0000	0.0002	0.0011	0.0040	0.0109	0.0240	0.0453	0.0764	0.1179	0.1693

The corresponding analysis for the proper game of Risk is more complex since the attacker has three dice and the defender has two dice. As a consequence, on each move in the general case, the attacker can lose two armies or the defender can lose two armies or they can lose one each. As a result the Chapman–Kolmogorov equation becomes

$$P_{m,n} = \lambda P_{m,n-2} + (1-\lambda-\mu)P_{m-1,n-1} + \mu P_{m-2,n}$$

where λ and μ are rather different from the values in the simple case discussed above. This problem is left to the reader as Exercise 3 in the lecture on Probability.

Queuing

PROBLEMS RELATING TO QUEUING typically involve time. Thus we might have a queue of people waiting to buy a ticket. There might be one or more servers. New customers might arrive at random and the time that servers take to deal with a customer might also be random.

By random here we will suppose that in a small interval of time Δt, the probability of a new customer arriving is $\lambda \times \Delta t$ and the probability of a customer finishing being served and so leaving is $\mu \times \Delta t$. If there is only one server and λ is greater than μ then the queue will grow in an unbounded manner. If λ is less than μ then the queue will not get out of hand and indeed there will be times when the queue is empty and the server is actually idle. It is convenient to denote the ratio λ/μ by r.

These arrival and serving processes are known as Poisson processes after the brilliant French mathematician Siméon Poisson (1781–1840).

Questions that might be asked are how long will the queue be on average and what fraction of time is the server idle? We can tackle such problems by using a similar approach to the Chapman–Kolmogorov equations. We assume that there is only one server.

The probability of the length of the queue being n at a time $t + \Delta t$ will be equal to the sum of

the probability of the length being $n-1$ at time t multiplied by the probability of one person arriving,

the probability of the length being $n+1$ at time t multiplied by the probability of one person leaving,

the probability of the length being n at time t multiplied by the probability that no-one arrives or leaves.

We assume that the interval Δt is so small that we can neglect the possibility of more than one event occurring.

So adding these three probabilities together we get

$$P_n(t + \Delta t) = \lambda P_{n-1}(t) + \mu P_{n+1}(t) + (1-\lambda-\mu)P_n(t)$$

where $P_n(t)$ denotes the probability of there being n people in the system (that is including a person being served, if any) at time t.

If we now also assume that the system is stable so that the probabilities do not vary with time, then $P_n(t + \Delta t)$ will be the same as $P_n(t)$ so that

$$(\lambda+\mu)P_n = \lambda P_{n-1} + \mu P_{n+1} \qquad n > 0$$

A special case arises if there is nobody being served (that is, the server is idle) since it is then not possible for someone to leave the system, so for $n = 0$ we have

$$\lambda P_0 = \mu P_1$$

Setting $r = \lambda/\mu$ we find that

$$P_1 = rP_0$$
$$P_2 = (r+1)P_1 - rP_0 = r^2P_0$$
$$P_3 = (r+1)P_2 - rP_1 = r^3P_0 \qquad \text{and so on}$$

However, since the sum of all the Ps must be 1, it follows that

$$P_0 \times (1 + r + r^2 + r^3 + \cdots) = 1$$

Now remembering that $1 + r + r^2 + r^3 + \cdots$ is equal to $1/(1-r)$ we find that

$$P_0 = 1-r \qquad \text{and} \qquad P_n = r^n \times (1-r)$$

So the probability of the server being idle is just $1-r$. The average queue length (counting anyone being served as really in the queue so far as new customers are concerned) is clearly

$$L = (0P_0 + 1P_1 + 2P_2 + 3P_3 + \cdots) / (P_0 + P_1 + P_2 + P_3 + \cdots)$$

Now the sum of the Ps is 1 so this becomes

$L = P_0(r + 2r^2 + 3r^3 + \cdots)$ and then multiply by r
$rL = P_0(r^2 + 2r^3 + 3r^4 + \cdots)$ subtract these and put $P_0 = 1-r$
$(1-r)L = (1-r)(r + r^2 + r^3 + \cdots)$ so cancelling $1-r$ we finally get
$L = r/(1-r)$

However, perhaps of more psychological interest is the average visible queue length (that is excluding anyone being served) and in a similar way we can easily compute that to be $Q = r^2/(1-r)$.

ratio r	0.95	0.9	0.85	0.8	0.75	0.7
per cent server idle P_0	5	10	15	20	25	30
mean visible queue Q	18.05	8.10	4.82	3.20	2.25	1.63

Idleness of server versus average queue length.

If we decide that the average queue length that customers will tolerate is perhaps 2, then we find that

$$r^2 = 2(1-r) \quad \text{so} \quad r^2 + 2r - 2 = 0$$

Solving this quadratic equation in the usual way gives

$$r = \sqrt{3} - 1 = 0.732...$$

from which it follows that $P_0 = 1 - r = 2 - \sqrt{3} = 0.2679...$. So in order to restrict the average queue size to be 2, we find that as a consequence the server is idle for over 26% of the time.

If on the other hand we don't want the server to be idle for more than 10% of the time then we find that $r = 0.9$ and the average queue length is 8.1. Gosh that will make the customers grumpy! So considerable idleness has to be accepted if the customers are to be kept happy. The table above shows the values of P_0 and Q for different values of the ratio r.

If the cost to society of the server being idle and the customers having to wait is the same per person at £x per hour then the overall cost C is clearly

$$C = P_0 x + Qx = x \left[(1-r) + r^2/(1-r)\right]$$

This has a minimum when $r = 1 - \sqrt{2}/2 = 0.29...$. In other words the server has to be idle some 70% of the time and the average queue length is then about 0.12. Tell that to your local railway station!

Further reading

A CLASSIC TEXT on this and related topics is Volume 1 of *An Introduction to Probability Theory and its Applications* by William Feller.

D Polydivisibility

IN THIS APPENDIX we look at an intriguing puzzle which provides lots of opportunities for using the various rules for divisibility. It can also be extended to use different bases and thereby really give the reader a headache.

The problem

ARRANGE THE NUMBERS 0 to 9 to form a ten-digit number such that the first n digits are divisible by n for all values of n from 1 to 10. For example consider

>   1234567890

The first digit (1) is of course divisible by 1 and so clearly any digit can be first. The first two digits (12) are divisible by 2 and so we conclude that the second digit must be even. The first three digits (123) are divisible by 3, but the first four digits (1234) are not divisible by 4 since $1234 \div 4$ is 308 remainder 2. So the number fails.

Before going further we remind ourselves of the relevant divisibility rules

2   Last digit must be even.
3   Digits must add to a multiple of 3.
4   Last two digits must be divisible by 4.
5   Last digit must be 0 or 5.
6   Must be divisible by both 2 and 3.
7   Do it the long way or use the 1001 rule in the lecture on Notations.
8   Last 3 digits must be divisible by 8.
9   Digits must add to a multiple of 9.
10 Last digit must be zero.

We rapidly see that since the ten digit number as a whole must be divisible by ten, the last digit must be 0. So the fifth digit must be 5 since we have just excluded 0. Also the digits in second, fourth, sixth, and eighth places must all be even; that uses up all the even digits and so the odd digits go in the odd places.

Another point is that since the sum of the digits from 1 to 9 is 45 which divides by 9, the ninth digit will automatically be acceptable, so we needn't worry about the ninth digit.

The obvious approach now is to consider all possible starts of three digits obeying the above rules. They are

   123, 129, 147, 183, 189
   321, 327, 369, 381, 387
   723, 729, 741, 783, 789
   921, 927, 963, 981, 987

That's twenty possibilities which seems a bit tedious. But we can simplify matters by considering the even digits.

The test for divisibility by 4 is that the last two digits must be divisible by 4; that is digits 3 and 4 must form a 2-digit number divisible by 4. Since digit 3 must be odd, we find that the only possibilities are

   12, 16, 32, 36, 72, 76, 92, 96

and we see that digit 4 can only be 2 or 6.

Now consider the first six digits. They must be divisible by 3 and by 2. We know that the first three digits will be divisible by 3 and so digits 456 must themselves be divisible by 3 as well. Since we know that digit 4 can only be 2 or 6 and digit 5 has to be 5, the only possibilities for digits 456 are

   258 or 654

So digit 6 can only be 4 or 8.

Now consider digit 8. In order to divide by 8, the 678 group must be divisible by 8. Since digit 6 must be even we only have to ensure that digits 78 are divisible by 8 since 200 is divisible by 8 anyway. And we know that digit 7 is odd. The only possibilities for digits 78 are therefore

   16, 32, 72, 96

So digit 8 can only be 2 or 6 – just like digit 4. So digits 4 and 8 have to be 2 and 6 and digits 2 and 6 have to be 4 and 8. In particular, digit 2 can only be 4 or 8.

We can now go back to considering the possibilities for digits 123 and we find that the twenty possibilities we had earlier can be reduced to just ten thus

   147, 183, 189, 381, 387
   741, 783, 789, 981, 987

If we now combine these with the fact that digits 456 can only be 258 or 654, we find that the only possibilities for the first six digits are

   147258, 183654, 189654, 381654, 387654
   741258, 783654, 789654, 981654, 987654

And now merging these with the only possibilities for digits 78 and adding the one remaining unused digit for digit 9 and the zero for digit 10 we end up with the following list of ten possibilities

1472589630, 1836547290, 1896543270, 1896547230, 3816547290
7412589630, 7896543210, 9816543270, 9816547230, 9876543210

And now we still have to check for divisibility by 7. Perhaps the simplest thing to do is use brute force. That eliminates all of them except one which is

3816547290

So that's it! It is easy to remember this answer. The even digits are backwards 86420 and the odd digits are forwards except that 1 and 3 are interchanged thus 31579.

It might be amusing to try the 1001 divisibility test for 7 by grouping the first seven digits into threes and then adding and subtracting the groups. Thus for the first one we get 1'472'589 giving

589 − 472 + 1 = 118.

and 118 divided by 7 has remainder 6 and so the original number 1472589 does not divide by 7.

Trying this on the other nine candidates we get the final set of ten numbers

118, −288, −352, −348, −266
184, −346, −264, −260, −324

and of these only −266 is divisible by 7.

Other bases

WE CAN DO this problem using other bases. For example consider octal (base 8). The problem then becomes to find arrangements of the digits 12345670 to form an eight-digit number in base 8 such that the first n in base 8 are divisible by n for all values of n from 1 to 7.

The divisibility rules for base 8 are

2   Last digit must be even.
3   Alternate digits added and subtracted must form a multiple of 3.
4   Last digit must be 0 or 4.
5   Do it the long way.
6   Must be divisible by both 2 and 3.
7   Digits must add to a multiple of 7.

The rules for 3 and 7 in base 8 (octal) were explained in the lecture on Notations. They are similar to the rules for 11 and 9 in normal decimal (base 10) notation. The rule for 4 in octal is similar that for 5 in decimal. And rules for 5 in octal are much like those for 7 in decimal – nasty.

We can quickly deduce that the last digit must be zero and digit 4 must be 4 just as digit 5 in the decimal problem had to be 5. And similarly we can deduce that the even digits must be even and the odd ones must be odd. Indeed, since digit 4 has to be 4, digits 2 and 6 can only be 2 and 6, or 6 and 2. And finally, since all the digits add to 28 the first seven digits will always pass the divisibility by 7 test.

To find the possibilities for the first three digits we have to find numbers such that the first plus last digit minus the middle digit is a multiple of 3. With a bit of care we have the following possibilities

127, 325, 523, 721
165, 561, 567, 765

Now consider digits 456. We know that digit 4 is always 4 and digit 6 can only be 2 or 6. If the first 6 digits are going to be divisible by 6 then they must be divisible by 3. Moreover, we can assume as before that the first 3 are divisible by 3 so digits 456 must also be divisible by 3. With care using the same rule we find the only possibilities are

432, 416, 476

Combining these two sets we find that the only possibilities for the first six digits are

325416, 325476, 523416, 523476
165432, 561432, 567432, 765432

We now have to check the first five digits for divisibility by 5. There are various tedious ways of doing this. We can convert the number to decimal and then the test is trivial, or we can do the division using octal arithmetic which needs care. We find that there are in fact three solutions. And then adding the last two digits we finally get the solutions to be

32541670, 52347610, 56743210

We leave the reader to explore other bases and in particular to show that it never works for odd bases. In base 4, the solutions are 1230 and 3210. In base 6, the solutions are 143250 and 543210. Now try base 12.

For further information on this curious topic and various variations see http//en.wikipedia.org/wiki/Polydivisible_numbers.

E Groups

THE THEORY OF GROUPS was largely developed in the nineteenth century and is one of the most important foundation concepts in modern mathematics. In some ways it bridges the gap between geometry and numbers. In the lecture on Bells we encountered various permutations as examples of groups and also mentioned the concept of cosets. This appendix gives a brief summary of some simple features of groups.

Basics

A GROUP consists of a set of elements (things, items, whatever) and an operation which can be performed on (that is, between) any two of the elements. The result of the operation has to be an element of the group.

Moreover, there has to be one element of the group (we call it the unit element and usually denote it by i) such that the result of applying the operation between it and any other element just gives that other element. Finally, to every element there has to be an element such that applying the operation between them gives the unit element. We say that one element is the inverse of the other.

As a simple numerical example the set of numbers 0, 1, 2, and 3 form a group if the operation is addition modulo 4. This is because the result of adding any two of the numbers modulo 4 gives an element of the set. There is a unit element (in this case 0) and every element has an inverse: the inverse of 1 is 3 and vice versa and the inverse of 2 is also 2.

Another important property of many groups is that the order of the elements when the operation is applied does not matter. Thus 1+2 is the same as 2+1. In such a case we say that the group is Abelian after the Norwegian mathematician Niels Abel (1802–1829).

Now consider the rotation of a square within its plane. We can denote the rotation of the square by a quarter turn by q, the rotation by a half turn by h and the rotation by a three-quarter turn by t. Also we can denote no rotation by i. These four rotations form a group where the operation is "followed by" and the unit element is i. If we denote the operation by $\times$ then we find that $q \times h = t$; that is a quarter turn followed by a half turn is the same as a three-quarter turn and so on. Clearly $q \times h$ is the same as $h \times q$ so that this group is Abelian. Again every element has an inverse; the inverse of q is t and the inverse of h is also h.

It should be clear that this group of rotations of the square and the group of additions modulo 4 are just different physical manifestations of the same abstract group. It is known as the Cyclic group of order 4 and is usually denoted by C_4.

It is more conventional to consider the quarter turn as the rotation r, the half turn as $r \times r$ or r^2, and the three quarter turn as r^3. In other words the group consists of the four elements i, r, r^2, r^3. Note that $r^4 = i$, the unit element.

We can have cyclic groups of any order. Thus the group of rotations of a triangle is C_3 with the three elements i, r, r^2. And of course in this case $r^3 = i$.

Subgroups

SOMETIMES A SUBSET of elements of a group forms a group in its own right. Thus in the case of the rotation of a square the identity and half turn form a group which in fact is C_2, the cyclic group of order 2. We say that C_2 is a subgroup of C_4.

It is clear that the general cyclic group C_n has subgroups corresponding to the various factors of n. Naturally, if n is a prime number then there are no proper subgroups at all.

The term *order* is important. The order of a group is simply the number of elements in it. The order of an element e is the smallest value of n such that e^n is the identity i. So in the case of the rotations of the square, the order of the group is 4, the order of the quarter turn is 4, the order of the half turn is 2, and the order of the three-quarter turn is 4. It is easy to show that the order of any element of a group has to be a factor of the order of the group itself. Similarly, the order of a subgroup is also always a factor of the order of the group.

Generators

WE HAVE JUST SEEN that the elements of the Cyclic group C_4 are i, r, r^2, r^3. We say that r is a generator of the group since the elements of the group are all simply powers of r. Note that r^3 is also a generator of the group.

Now consider the manipulation of a square when we also add the ability to flip it over and for simplicity we consider just flipping it about the vertical axis. This was mentioned in the lecture on Bells when discussing rounds and plain hunting. If the operation of a quarter turn is r as before and the operation of doing a flip is f, then we see that doing f then r (that is fr) is not the same as doing r then f. In other words they do not commute and the group is not Abelian.

This group of the positions of a square is known as D_4, the Dihedral group of order 4. It has 8 elements which are $i, r, r^2, r^3, f, fr, fr^2, fr^3$. These are illustrated in the diagram opposite. From the diagram we easily see that rf is in fact the same as fr^3.

So we see that D_4 has two generators, r and f where the order of r is 4 but the order of f is 2. (Two flips cancel out.) Moreover, r and f are related by the fact that $(rf)^2 = i$ and by equations such as $rf = fr^3$.

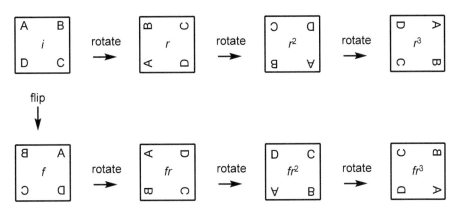

The elements of D_4, the Dihedral group of order 4.

Note also that D_4 has a number of subgroups. Clearly i, r, r^2, and r^3 form a subgroup of order 4 (in fact a C_4). And i and f form a subgroup of order 2 (a C_2). Also i and r^2 form a subgroup of order 2 (another C_2). In addition, the elements fr, fr^2, and fr^3 all have order 2 (just try doing them twice) so with i they all form subgroups of order 2. Curiously, the only elements of order 4 are r and r^3.

Another way of arranging the diagram is shown below in which the vertices represent the elements and the lines represent the operations. The lines with arrows denote r and those without denote f. Since f has order 2 it can be done either way and needs no arrow. This style of presentation of a group by a graph was introduced by the English mathematician Arthur Cayley (1821–1895). Cayley was Senior Wrangler in 1842 and like Colenso mentioned in the lecture on Measures was also one of Mr Hopkins' men. By tracing the routes on this diagram it is easy to see that rf is the same as fr^3 and that $(fr)^2 = i$.

We encountered similar graphs in the lecture on Bells. The graph for the allowed changes on three bells was just a hexagon, whereas that for the allowed changes on four bells was an elaborate structure consisting of linked squares and hexagons on a projective plane!

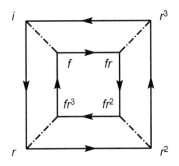

The Cayley graph of D_4.

Cosets

WE NOW CONSIDER the important concept of a coset which has been briefly mentioned in the context of bell ringing and will occur again when we look at the arrangements of a Rubik cube in Appendix F.

The idea is to take a subgroup and consider the set formed by applying the group operation between some fixed element e and each element of the subgroup in turn. As an example, consider the group D_4 and the subgroup H consisting of (i, r, r^2, r^3).

If the fixed element e is r then the result is a set we can call $r \times H$ consisting of $(r \times i, r \times r, r \times r^2, r \times r^3)$ which is of course (r, r^2, r^3, i). This is the set H again although the elements are given in a different order. We will inevitably get the set H again if e is a member of H in the first place.

But suppose we choose e not to be in H; take f as an example, then we get the set consisting of $(f \times i, f \times r, f \times r^2, f \times r^3)$, that is the set (f, fr, fr^2, fr^3). This set is not a subgroup because it does not include i for one thing. It is called a coset and can be denoted by $f \times H$. If we do the operation in reverse then we get the set consisting of $(i \times f, r \times f, r^2 \times f, r^3 \times f)$ which actually is the set (f, fr^3, fr^2, fr). This coset can be denoted by $H \times f$ and is in fact the same as $f \times H$ although the elements are given in a different order. So in this example the left coset $f \times H$ and right coset $H \times f$ are the same but that is not always so. A key point is that the group as a whole is subdivided into the subgroup and a coset. This subdivision is illustrated below. The subgroup elements are framed in black and the coset elements in red.

Remember that D_4 has a number of subgroups of order 2. We will now look at their associated cosets.

We start by considering the subgroup consisting of just (i, r^2); we will call it K. Now if the fixed element e is itself r^2 then clearly the set $e \times K$ is just K once more. If e is r then we get $r \times K$ which is the coset (r, r^3); if e is r^3 we get the coset (r^3, r) which is the same. If e is f then we get $f \times K$ which is the coset (f, fr^2). Similarly if e is fr we get the coset (fr, fr^3). If e is fr^2 or fr^3 we get the same cosets once more. So now we see that the group is subdivided into the subgroup K and three cosets. This subdivision is also shown below with the subgroup in black and the three cosets in various colours.

If we consider right cosets $K \times e$ then the cosets are the same and so we get the same decomposition.

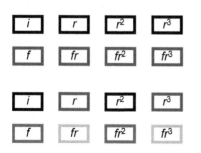

The group D_4, divided into the subgroup $H = (i, r, r^2, r^3)$ of order 4 and a coset of order 4.

The group D_4, divided into the subgroup $K = (i, r^2)$ of order 2 and three cosets of order 2.

E Groups 253

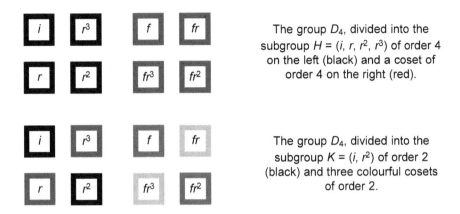

The group D_4, divided into the subgroup $H = (i, r, r^2, r^3)$ of order 4 on the left (black) and a coset of order 4 on the right (red).

The group D_4, divided into the subgroup $K = (i, r^2)$ of order 2 (black) and three colourful cosets of order 2.

Another way of representing the cosets is to note that the Cayley graph of D_4 shown earlier looks exactly like a representation of a cube. We imagine that each of the eight corners has a small cube denoting the eight elements. And then we can put the two layers of the cube side by side so the subdivisions by the subgroups H and K can be represented as shown above.

We will now look at the other order 2 subgroups. Thus consider the subgroup consisting of just (i, f); we will call it L. The left coset $r \times L$ consists of the two elements (r, rf); but rf is actually fr^3. And the right coset $L \times r$ consists of (r, fr). So in this case the left and right cosets are different. In the diagram below these cosets are shown in red.

We will persevere with the left cosets, as we just saw $r \times L$ is (r, fr^3) in red below. Similarly $r^2 \times L$ is (r^2, fr^2) in blue and $r^3 \times L$ is (r^3, fr) in green.

Pressing on we have $f \times L$ is (f, i) because f^2 is i so in this case the coset becomes simply the subgroup L (black); now $fr \times L$ is (fr, frf) but frf is in fact just r^3 so we have (fr, r^3) (green). In the same way $fr^2 \times L$ becomes the subgroup L (black) and $fr^3 \times L$ becomes (fr^3, r) (red).

Well that was all rather tedious but the outcome is depicted below left. If we do the same with right cosets we get the slightly different decomposition shown below right in which fr and fr^3 have changed over. Note that the left and right blue cosets which involve r^2 are the same, but the red and green cosets which involve r and r^3 are different. It is clear that r and r^3 are awkward.

The group D_4, divided into the subgroup $L = (i, f)$ of order 2 and three cosets of order 2. In this case the left and right cosets are different.

254 Nice Numbers

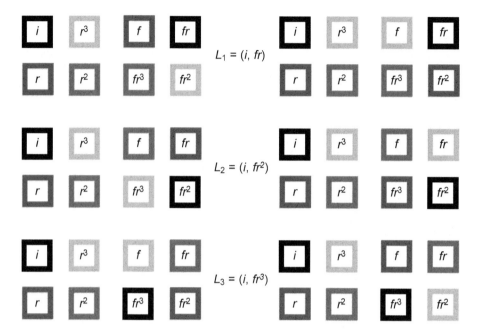

$L_1 = (i, fr)$

$L_2 = (i, fr^2)$

$L_3 = (i, fr^3)$

We should also explore the cosets of the other order 2 subgroups which are (i, fr), (i, fr^2), and (i, fr^3) which we can call L_1, L_2, and L_3 respectively. The results are shown above. In each case the left and right cosets are also different.

If the left and right cosets of a subgroup are the same then the subgroup is said to be a normal subgroup. So in this example, H and K are normal subgroups but L, L_1, L_2, and L_3 are not normal subgroups.

Permutations

IN THE LECTURE on Bells, we saw that the various changes were simply represented by permutations of numbers. Thus in the case of three bells the changes were 123, 213, and so on. These changes form a group traditionally known as S_3. The unit element is 123 and is the state when ringing rounds.

The group S_3 has 6 elements which can be rung in the order 123, 213, 231, 321, 312, 132. Permutations are known as odd or even according to whether an odd or even number of single transpositions are required to produce them. So 123, 231, and 312 are even and 213, 321, and 132 are odd. The even ones form a subgroup of order 3 and the odd ones form the corresponding coset.

The Cayley graph of S_3 is shown opposite. It has exactly the same form as the graph for the Dihedral group D_3 which describes rotating and flipping a triangle in the same way that D_4 describes the positions of a square. Thus we have operations r and f with $r^3 = f^2 = (rf)^2 = i$. The outer triangle gives the subgroup of even permutations and the inner triangle is the coset of odd permutations. It is perhaps best to consider the graph to be a triangular prism that is projected onto the plane for convenience.

We say that S_3 and D_3 are isomorphic from the Greek ισo (*iso*), the same and μορφη (*morfh*), form. So to all intents and purposes they are the same group. The flip of D_3 corresponds to the operation (1, 3) on the bells which we recall is not allowed. The operations which are allowed are

a (1, 2)   interchange first two

b (2, 3)   interchange last two

It is easy to see that a corresponds to fr and that b corresponds to fr^2. Moreover, fr^2 is the same as rf. (In much the same way as fr^3 is the same as rf in D_4.) So because of the constraints on which interchanges of bells are permitted, alternate moves correspond to fr and rf and go via the diagonals of the rectangles of the triangular prism as shown in red and blue. This twisted route corresponds to the simple hexagonal graph in the lecture on Bells!

In the case of four moving bells we encounter the group S_4 with 24 elements. When ringing Plain Bob Minimus we noted that doing alternate changes a and b would return us to rounds after 8 changes and that we had to insert changes c to do the full course. This is because changes a and b generate only a subgroup of S_4. This subgroup has two cosets and doing change c shifts us from the subgroup to one coset and then to the other. We will look at the Cayley graph of S_4 in a moment.

Similarly, in ringing Plain Bob Doubles with five moving bells we encounter the group S_5 with 120 elements. Here we have two levels of subgroups and cosets. There is a subgroup with 40 elements and two corresponding cosets; we switch from one to the other whenever a Bob is performed. And within the Plain Course of 40 elements we have a subgroup of 10 changes with three cosets and the change c moves us between them. And so on ....

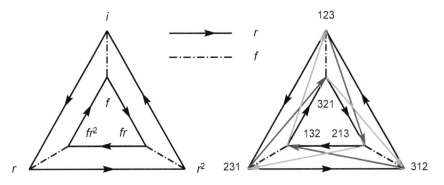

The Cayley graphs of D_3 and S_3.   The changes on three bells.

Polyhedral groups

THERE ARE many other forms of groups. Taking the even permutations of a permutation group S_n gives the Alternating group A_n. Thus the even permutations of four items are described by A_4. The number of elements in A_n is of course $n!/2$ which in the case of A_4 is 12.

We have seen that the possible positions of a polygon are described by C_n or D_n if we allow flipping. The polyhedra are more complex. The possible positions of a tetrahedron are described by A_4; those of a cube or octahedron by S_4; and those of a dodecahedron or icosahedron by A_5.

In the case of a tetrahedron, we can choose the base triangle in four ways; we can then rotate the base into three positions (it is C_3 of course); hence we see that the tetrahedron can be positioned in 12 ways which is consistent with the number of elements in A_4. In the case of a cube we can choose the base in six ways and rotate it into four positions giving 24 in total (as for S_4) and for the dodecahedron we have 12 times 5 giving 60 positions (as for A_5).

We will now look at the tetrahedron in more detail. We can label three vertices as shown below with ABC being the base and then D is the other vertex. We define two operations, one is the rotation about the axis joining D to the midpoint of the base ABC which we denote by r. Thus applying r to the tetrahedron ABCD rotates it to produce CABD. The other operation is a half turn about the axis joining the midpoints of AB and CD which we denote by f (for flip). So applying f to ABCD interchanges the pairs A and B and also C and D thus producing BADC.

Thus we have

f	(1, 2)(3, 4)	interchange two pairs	$f^2 = i$
r	(1, 2, 3)	rotate first three	$r^3 = i$

The Cayley graph for the tetrahedral group is a truncated tetrahedron as shown below. The starting position ABCD is taken as the unit element. The

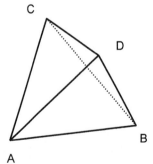 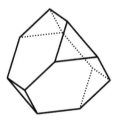

A tetrahedron and the Cayley graph for A_4.

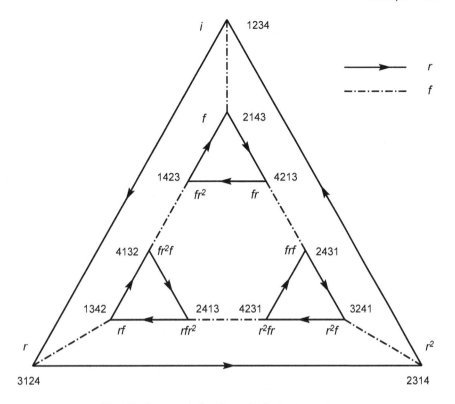

The Cayley graph for A_4 projected onto a plane.

triangle corresponding to the vertex D represents the elements ABCD, BCAD, and CABD. It is a bit easier to understand if the truncated tetrahedron is projected onto a plane as shown above. The elements have been numbered as 1234 rather than ABCD to make comparison with the bells more straightforward. The corresponding description as operations such as rf are shown as well.

Note that r and f are related since applying rf three times brings us back to the beginning so that $(rf)^3 = i$. Also $(fr)^3 = i$ as well. These correspond to going around a hexagonal face of the truncated tetrahedron. Thus the position marked fr^2f is also rfr and frf is also r^2fr^2.

Looking at the Cayley graph it is obvious that it is A_4 because it only shows the even permutations of 1234. The odd permutations would correspond to the tetrahedron being reflected as in a mirror and we cannot do that without going into the fourth dimension. Note also that the operations r and f are both even permutations and so it is not possible to generate an odd permutation anyway.

We now turn to consider the transformations of a cube. At first sight this is a bit more difficult. The cube has eight vertices but our goal is to represent its position by simply all the permutations of ABCD, so it is clear that it is not

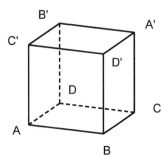

The cube has four space diagonals AA', BB', CC', and DD'.

sensible to consider labelling the eight vertices as ABCDEFGH. The secret is to label the vertices of one face as ABCD and the opposite face as A'B'C'D' so that the lines AA', BB', CC', and DD' are the four space diagonals of the cube as shown above. This has the consequence that every face has a combination of ABCD where we simply ignore the primes.

We define two operations, one is rotation about a vertical axis so that the base ABCD becomes DABC which we call r. The other operation is a half turn about the axis joining the midpoints of AB and A'B' which we donate by f (flip once more). Applying f to ABCD just interchanges A and B to give BACD (remember that we are ignoring the primes; what we are really doing is manipulating the four space diagonals).

Thus we have

f	(1, 2)	interchange first two	$f^2 = i$
r	(1, 2, 3, 4)	rotate four	$r^4 = i$

Note the close similarity to the operations on the tetrahedron. The Cayley graph for the cubic group is in fact a truncated octahedron which is shown below. Again we project the figure onto a plane as shown opposite.

The elements are again numbered as 1234 rather than ABCD so that comparison with bells is easier.

As in the case of the tetrahedron we have $(rf)^3 = i = (fr)^3$ and these correspond to going around a hexagonal face of the truncated octahedron. Again we have relationships such as $fr^2fr^2 = r^2fr^2f$.

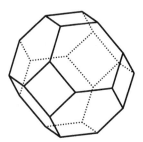

The Cayley graph for the cubic group S_4 is a truncated octahedron.

E Groups 259

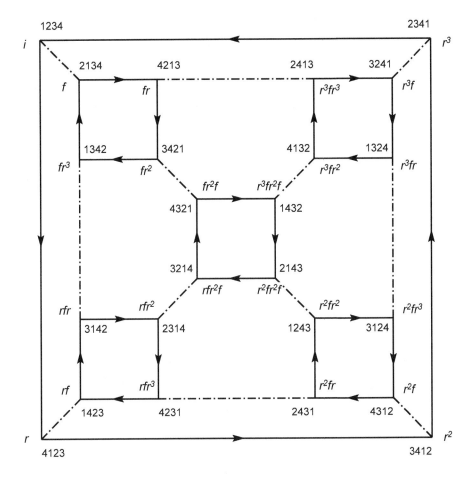

The Cayley graph for S_4 projected onto a plane.

We can now perhaps consider how ringing the changes on four bells corresponds to moves on the Cayley graph. Recall the following changes

- a (1, 2)(3, 4)   interchange two pairs – a double
- b (2, 3)   interchange 2 and 3
- c (3, 4)   interchange 3 and 4
- d (1, 2)   interchange 1 and 2.

The double a corresponds to going to the opposite vertex of the truncated octahedron and is fr^2fr^2. The move b is r^3fr, the move c is r^2fr^2, and the move d is simply f. We leave to the reader the task of tracing some of these moves on a model of a truncated octahedron. It's probably not worth the effort!

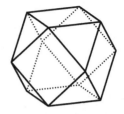 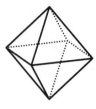

The cube, cuboctahedron, and octahedron all have the same group properties.

We have noted that the cubic group S_4 has elements of order two (f) and four (r). But it also has elements of order three such as (rf) which correspond to rotation about the space diagonals. The dual of the cube is the octahedron and has identical group properties. An important figure is the cuboctahedron which is an intermediate figure and also has the same group properties; they are all shown above. The cuboctahedron has both square and triangular faces and so clearly reveals the operations of order three and four; the operations of order two are simply the rotations about its space diagonals.

The two other regular polyhedra are the dodecahedron with 12 faces and 30 vertices and its dual, the icosahedron, with 30 faces and 12 vertices. In this case the transformations are described by the group A_5 which has 60 elements and the properties are similar to those of the tetrahedron and cube but we will not go into them in detail. Suffice it to say that we can define r (corresponding to a rotation of a pentagonal face or vertex) and f (a flip) with $r^5 = i$ and $f^2 = i$. Moreover, we also have $(rf)^3 = i$ as in the case of the tetrahedron and cube. The Cayley graph is a truncated icosahedron which has pentagonal and hexagonal faces (exactly as a soccer ball).

The vital statistics of the groups of the polygons and polyhedra that we have looked at are given in the table below.

We have noted that the order of a subgroup is always a factor of the order of the group. This was discovered by the French mathematician Joseph-Louis Lagrange (1736–1813). The converse is not quite true. However, if the order of a group has a factor which is the power of a prime, then it will have a subgroup of that order. Thus the tetrahedral group (which has order 12) has subgroups of orders 2, 3, and 4 but not 6. This was discovered by the Norwegian mathematician Ludwig Sylow (1832–1918).

Object	Group	Order	Generators	Graph
Triangle	D_3	6	$f^2 = r^3 = (fr)^2 = i$	triangular prism
Square	D_4	8	$f^2 = r^4 = (fr)^2 = i$	square prism
Tetrahedron	A_4	12	$f^2 = r^3 = (fr)^3 = i$	trunc tetrahedron
Cube	S_4	24	$f^2 = r^4 = (fr)^3 = i$	trunc octahedron
Dodecahedron	A_5	60	$f^2 = r^5 = (fr)^3 = i$	trunc icosahedron

Direct products

WE HAVE NEGLECTED to mention one of the simplest groups of all. This is D_2 and is often known as the four group. Remember that D_3 and D_4 concern rotating and flipping a triangle and square. The best analogy for D_2 is to consider doing the same to a ruler as shown below. We assume the ruler is transparent.

We can call the operations r and f as before but the big difference is that they both have order 2. The Cayley graph is simply a square as also shown below. A very important point is that D_2 is an Abelian group unlike the other dihedral groups and so $fr = rf$.

It is interesting to compare D_2 with C_4, the cyclic group which also has four elements. A good way to do this is to show their multiplication tables.

C_4	i	r	r^2	r^3
i	i	r	r^2	r^3
r	r	r^2	r^3	i
r^2	r^2	r^3	i	r
r^3	r^3	i	r	r^2

D_2	i	r	f	fr
i	i	r	f	fr
r	r	i	fr	f
f	f	fr	i	r
fr	fr	f	r	i

On the left is the table for C_4 and on the right is the table for D_2. They should be self-explanatory. Each entry is the result of taking the first operand from the column on the left and the second from the row at the top. Note especially the structure of the table for D_2. It consists of four subtables which are in pairs. The first subtable has entries i, r, r, i and the second has entries which are these multiplied by f.

In fact D_2 is really a combination of two cyclic groups C_2, one for r and one for f. And without going into detail, we say that D_2 is the direct product of the two cyclic groups and we write $D_2 = C_2 \times C_2$. (Note that $D_1 = C_2$.)

We can form the direct product of any two groups provided that any element of one commutes with any element of the other. Interestingly, any Abelian group is the direct product of cyclic groups. We will encounter other examples when we consider the Rubik cube in Appendix F.

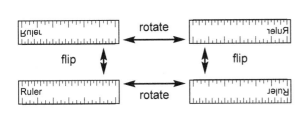

Rotating and flipping a ruler illustrate D_2.

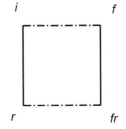

The Cayley graph for D_2.

The quaternion group

WE CONCLUDE by looking briefly at a strange group. In the case of complex numbers, i defines a rotation in the complex plane by 90 degrees (see the lecture on Primes). Clearly, 1, i, -1, $-i$ form an example of C_4. The Irish mathematician William Rowan Hamilton (1805–1865) struggled for some time to find a corresponding description of rotations in three dimensions. Eventually he came up with Quaternions which describe rotations in up to four dimensions.

As well as the complex unit i, there are also j and k with the following curious properties

$$i^2 = j^2 = k^2 = -1 \qquad i \times j \times k = -1$$
$$i \times j = k, \; j \times k = i, \; k \times i = j \qquad j \times i = -k, \; k \times j = -i, \; i \times k = -j$$

The group describing the quaternions has eight elements which are

$$1, \; -1, \; i, -i, \; j, \; -j, \; k, -k.$$

The quaternion group is strange. It is not Abelian (ij is not the same as ji) but all its subgroups are normal, that is their left and right cosets are identical.

An alternative way of looking at this group is to note that we can describe the elements just in terms of the unit and two of i, j, k. It is confusing that we normally use i for the unit whereas here it means a square root of minus one. So we will use capital I for the unit and rename two of i, j, k to be a and b so that the third is ab. To be specific suppose i is a and j is b so that k is ab. Note that

$$a^2 = b^2 = (ab)^2 \quad \text{and} \quad a^4 = b^4 = (ab)^4 = I$$

The eight elements in the order given before are now

$$I, \; a^2, \; a, \; a^3, \; b, \; b^3, \; ab, \; ba$$

and the Cayley graph is shown below.

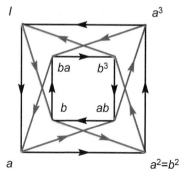

The Cayley graph of the quaternion group.

Further reading

A VERY READABLE INTRODUCTION to groups will be found in *Groups and their Graphs* by Grossman and Magnus. This covers many kinds of groups and illustrates their properties through the use of Cayley graphs.

Other examples of groups will be found in Appendix F on the Rubik cube.

F Rubik

FROM TIME TO TIME, the world is struck by a craze. There was the 15-puzzle in the 1880s, the Rubik cube in the 1980s, and now everybody fiddles around with Sudoku. In this appendix we look at the Rubik cube in some detail especially with regard to the illustration of various properties of groups.

Basics

THE CUBE was invented by Ernö Rubik in about 1974. It really took off in the 1980s. It faded in popularity but has recently seen a revival.

The cube comprises a 3 by 3 by 3 arrangement of small cubes. Actually only 26 because the centre one is basically the pivot. The faces, each of nine cubes, can be rotated. The cubes are coloured and in the normal (start) position the faces are typically white, red, yellow, orange, green, and blue. Usually white is opposite yellow, red opposite orange, and green opposite blue – the difference between the colours of the pairs of opposite faces is yellow in every case.

It is important to note that there are three kinds of little cubes

Centre cubes   these never move and identify the face colours. There are 6.

Edge cubes   these have two coloured faces. There are 12.

Corner cubes   these have three coloured faces. There are 8.

We denote the faces of the cube by L (left), R (right), U (up), D (down), F (front), and B (back). In the diagram below left, we can see the U, F, and R face. The colours of the other faces are D (white), L (orange), and B (green).

We also use the letters L, R, U, D, F, and B to denote clockwise quarter turns of the corresponding faces. If we hold the cube in our left hand with the blue face towards us and turn the right hand face (the face R) by a quarter turn away from us, we call this the movement R. This results in the cube shown below right.

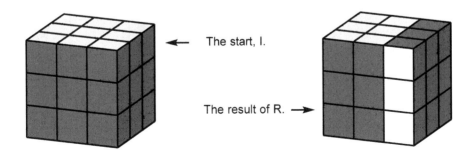

← The start, I.

The result of R. →

© Springer International Publishing Switzerland 2016
J. Barnes, *Nice Numbers*, DOI 10.1007/978-3-319-46831-0_16

A few casual twists of the faces reduce the cube to a mess; we will explain how to restore a chaotic cube to the starting point in a moment.

However, if the opposite faces are always turned in pairs, the resulting patterns are quite attractive being symmetric on each face. Returning such symmetric patterns to the start is relatively easy and it is wise for a new owner of a cube to experiment with such moves to start with.

Note that R^2 denotes a half turn and R^4 denotes a whole turn which is "no change" and is usually denoted by I (the Identity). A quarter turn anticlockwise is denoted by R^{-1} and a half turn anticlockwise by R^{-2}. The latter is of course also a half turn clockwise and so equally denoted by R^2.

If we turn the right slice by a half turn and then the left slice by a half turn then the total movement is denoted by R^2L^2. Similarly, a half turn of the front and back slices is F^2B^2 and of the up and down slices is U^2D^2.

If we denote the three moves R^2L^2, F^2B^2, and U^2D^2 by R, F, and U then there are 8 possible positions: I, R, F, U, FU, UR, RF, RFU. (Note that we often use the same symbol to represent an operation and the position obtained by applying that operation to the starting position I.) These operations form a group and are all shown below. Each operation has order 2 so doing it twice gets back to the initial position. Hence we write $F^2 = I$. Moreover, the operations all commute which means that the order of doing two operations is immaterial. So $RF = FR$. Thus the group is Abelian. It is usually referred to as the Slice-squared group.

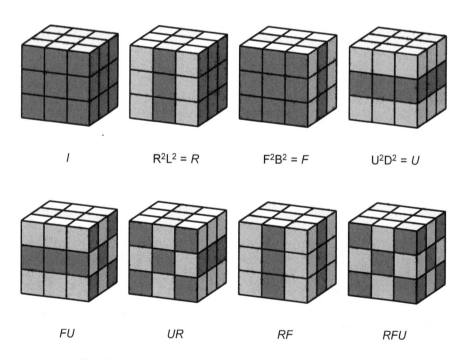

The Slice-squared group has 8 elements as shown above.

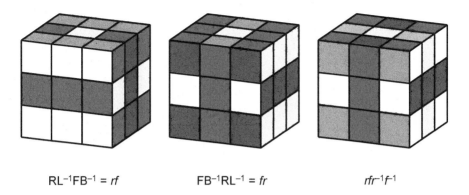

RL⁻¹FB⁻¹ = *rf*   FB⁻¹RL⁻¹ = *fr*   *rfr⁻¹f⁻¹*

A more interesting group is that using quarter turn slices such as RL⁻¹. The other basic moves are FB⁻¹ and UD⁻¹. For simplicity we can refer to these three moves as r, f, and u. (So $R = r^2$ and so on) We obviously have $r^4 = f^4 = u^4 = I$. So these operations have order 4. An important point to note about this group, which we can call the Slice group, is that it is not Abelian. Thus rf is not the same as fr as shown by the left and centre diagrams above. This is further emphasized by $rfr^{-1}f^{-1}$ at top right which would be I if the group were Abelian.

Note that we write I or I to mean the Identity and use italic for uniformity when dealing with moves that are written in italic such as r, f, and u.

It should be noticed that all the faces we have produced by doing slice moves take the form shown below. The corners are always the same and opposite edges are always the same. Up to four colours are possible.

In the case of the Slice-squared group, each face has a maximum of two colours. There are three possibilities: 1) c = a and d = b (stripes); 2) a = d and b = c (chessboard); 3) a = b = c = d (plain).

In the case of the more general Slice group, four different colours are possible. Thus the top face in the pattern $rfr^{-1}f^{-1}$ has a = blue, b = red, c = green, and d = yellow.

We saw above that the Slice-squared group has 8 elements (that is operations). Remember that the order of an element has to be a factor of the order of the group. So all operations of the Slice-squared group have to be factors of 8 and indeed they are all of order 2. The more general Slice group is much larger and has 768 elements. Note that $768 = 2^8 \times 3$. So in the case of the Slice group we might hope to find elements of orders 3 and 6.

a	b	a
c	d	c
a	b	a

General form of a slice face. Note that the corners are always the same. And opposite edges are always the same.

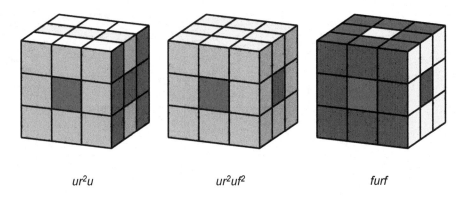

ur^2u ur^2uf^2 *furf*

Other interesting patterns are those in which a = b = c and d is different (spots). As an example consider the sequence ur^2u (that is uRu) which results in the pattern shown above top left. It is fairly easy to see the logic of these moves which are designed to change all the front face to green except for the centre which is of course fixed at blue. We know that u^2 would make the top and bottom rows of the front face have the required reverse colour green, but that does not help with the side edges which would remain blue. However, inserting r^2 between the two instances of u, interchanges the edges and so swaps the blue and green edges whereas the top and bottom rows (which are the same anyway) stay green. We get a familiar stripe on top whereas the right face (a big H) has a = b = d which we have not met before.

If we now do f^2 so that the total move is ur^2uf^2 we get the top centre pattern (four spots) which has a certain simplicity and is a good test for a novice to straighten out. The answer is simply to do ur^2uf^2 again and we find that we are back to I. So the operation ur^2uf^2 is another example of an operation of order 2.

Surprisingly, the simpler *furf* results in spots on all six faces as shown above top right. In this case the three spots are rearranged cyclically thus yellow → red → blue → yellow. If we do it again we find a different cyclic arrangement. If we do it a third time we get back to I. So this operation has order 3.

Another interesting group is the antislice group whose basic elements are RL, FB, and UD. We can denote these operations by r^*, f^*, and u^* by analogy with the slice operations r, f, and u.

The antislice group produces further pretty patterns. If we do $f^*u^{*-1}r^{*-1}f^*$ then we get the pattern shown opposite top left. It is a variation on six spots where the surround comprises two L-shaped blocks. Opposite faces have the complementary pattern. Thus the back is also orange and red but red is bottom right whereas the front has orange bottom right. Note that $f^*u^*r^*f^*$ does not produce the desired result (four faces are as desired but two are a bit of a tangle).

The centre pattern has zigzags on four faces. It is simply done by $(f^*r^*)^3$. A further application of u^{*2} (which is the same as u^2) leads to the right pattern in which the zigzags are replaced by z shapes.

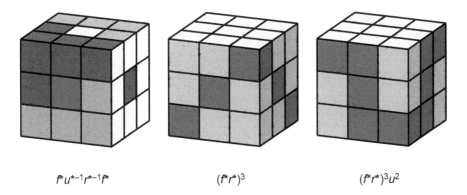

$f^*u^{*-1}r^{*-1}f^*$　　　　　$(f^*r^*)^3$　　　　　$(f^*r^*)^3u^2$

The operation F^2R^2 is interesting. If applied three times it will be found that two pairs of edge cubes in the top and bottom faces will be interchanged but the remainder of the cube is untouched. This operation is often useful for making corrections in the event of an error. The effect is shown below left.

Clearly F^2R^2 has order 6. Surprisingly, the related operations F^2R and FR have orders 30 and 105 respectively. And FR^{-1} has order 63. Note that LR simply has order 4 because L and R individually have order 4 and commute.

In order to get a feel for why these large orders occur, it is important to note that any move performs operations on both edges and corners. But these are independent in the sense that an edge cannot become a corner or vice versa. In the case of F^2R, applying it 6 times restores the corners whereas applying it 10 times restores the edges. Now the least common multiple (lcm) of 6 and 10 is 30 so it has to be applied 30 times to return to the start. Perhaps we can say its signature is (6, 10).

Similarly FR has signature (15, 7) giving lcm of 105 and FR^{-1} has signature (9, 7) giving lcm of 63. And F^2R^2 has signature (6, 3) with lcm 6. We will revisit this topic in a moment.

Another handy operation is $U^2R^{-1}LF^2RL^{-1}$ or $U^2r^{-1}F^2r$. This cycles three edge cubes of a slice. Two views of the result are shown below.

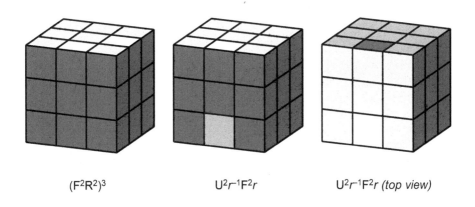

$(F^2R^2)^3$　　　　　$U^2r^{-1}F^2r$　　　　　$U^2r^{-1}F^2r$ *(top view)*

Restoring a cube

WE NOW consider the vital task of getting a randomly jumbled cube back to normal. We describe a foolproof method of doing this although an expert will spot many shortcuts. Beware that as few as four random moves will make the cube hard to unravel for normal mortals.

We now revert to using the basic operations R, L, F, B, U, D in contrast to the slice operations *R*, *F*, and *U* which were given in italic to distinguish them.

We need notations for the individual places. We use a pair of letters such as UF to mean the edge cube between the U and F faces and a triple such as FUR to mean the corner cube between the F, U, and R faces. See the diagram below.

The first goal is to get the bottom slice (white face down) correct. Although we normally work with the cube with white at the bottom, blue at the front, and red at the right, in order to correct the bottom slice it is convenient to invert the cube so that white is on top, red at the front, and blue at the right. Remember that the colours of the faces are defined by the central cubes and these do not move.

It is best to get the edges correct first and then tackle the corners. The four white edges could be initially in various places. On the bottom face or on a side face. And they can have different orientations. We consider these in turn.

Suppose the blue/white edge is on the bottom slice. Rotate the bottom slice until it is in the correct face (that is the right face). If it is oriented correctly (blue at the side) then all we need do is R^2. However, it might be that it is oriented with the white face at the side in which case do $RUF^{-1}U^{-1}$. The effect of this on a cube in the inverted starting position is shown opposite above left. The key thing is that DR has been moved to UR and has also been flipped. We write DR → RU.

If the blue/white edge is on the top edge but in the wrong place then simply rotate the face so that it is on the bottom slice, rotate the bottom slice as necessary, and then do as above.

If it is on a side edge such as FR then $R^{-1}D^{-1}R$ as shown opposite above centre will bring it to the bottom and not disturb the top edge. Then as before.

If it is on the top edge and in the right place but flipped then $R^{-1}UF^{-1}U^{-1}$ will flip it without disturbing the other top edges as opposite above right. In other words do R^2 to bring it to the bottom slice and then $RUF^{-1}U^{-1}$ as before.

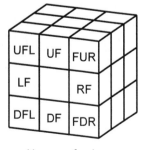

Names of cubes.

The cube inverted

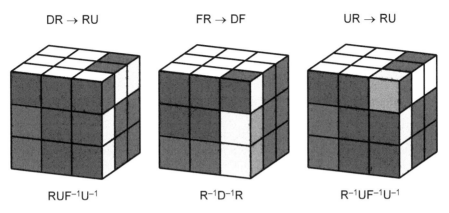

Operations on the white edges.

These simple moves enable the top edges to be put in the correct place. Now we have to tackle the corners without messing up the edges. The idea is to get the corner piece on the bottom slice and below where we want it. We then need three moves corresponding to three possible orientations.

If the white face is on the right then we need to move FRD to FUR. This can be done by $R^{-1}D^{-1}RD$ as shown below left. It also moves FUR back to DFR (note the twist) and similarly interchanges BDL and BRD with a twist. If the white face is at the front then we need to move DFR to FUR. The inverse $D^{-1}R^{-1}DR$ does just this as shown below middle. These moves are of order 6.

If the white face is at the bottom then we need to move RDF to FUR and this can be done by $FD^2F^{-1} D^2 R^{-1}DR$ as shown bottom right. These moves suffice to sort out the corners and thus get the top slice correct.

If we end up with a corner piece in the correct place but wrongly oriented, then we simply move it to the bottom slice and try again.

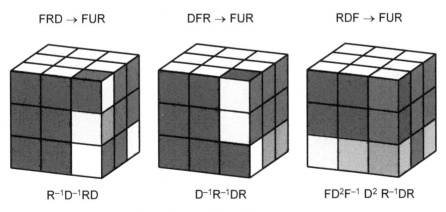

Operations on the white corners.

We now invert the cube so that the white face is at the bottom. This means that the bottom slice is now correct. Our next task is to correct the middle slice. Of course the centre cubes are always correct so that all we can need to do is to correct the edges of the middle slice.

Start by putting the edge on the middle slice between the front and right faces into place. Suppose it is somewhere on the top slice. Turn the top slice so that it matches the front or right face as appropriate. It will now be at UF or UR.

Consider the operation $URU^{-1}R^{-1}U^{-1}F^{-1}UF$ and apply it to the cube in the starting position. The result is shown in the middle below. The key purpose of this operation is to move the cube UF to RF. Moreover, the bottom slice has not been changed at all and no other changes have been made to the middle slice. The cube that was at RF is now at BU but that is of no interest. We write the move as

UF → RF : $URU^{-1}R^{-1}U^{-1}F^{-1}UF$

Now consider the operation $U^{-1}F^{-1}UFURU^{-1}R^{-1}$ and apply it to the cube in the starting position. The result is shown below right. This has moved the cube UR to FR. Again the bottom slice is unchanged and no other changes have been made to the middle slice. We write this move as

UR → FR : $U^{-1}F^{-1}UFURU^{-1}R^{-1}$

which is largely the reflection of the move for UF → RF.

If the cube we want is in the right place but flipped then we can simply move some other cube on the top edge into its place and then move it back with the correct orientation.

So assuming that our cube now has the middle and bottom slices correct we need to sort out the top slice. We do this in several stages. The first is to consider the edge pieces. Not only are they likely to be in the wrong place but they may be incorrectly oriented with the colour of the top at a side and so need to be

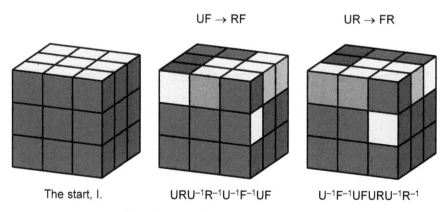

UF → RF   UR → FR

The start, I. $URU^{-1}R^{-1}U^{-1}F^{-1}UF$ $U^{-1}F^{-1}UFURU^{-1}R^{-1}$

Operations on the middle slice edges.

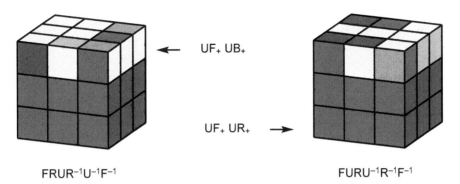

Flipping the top slice edges.

flipped before we do anything else. It might be that none need flipping or two need flipping or maybe they all need flipping. There will never be an odd number. The following moves do this flipping The first applies if the two pieces to be flipped are on opposite edges, the second if they are on adjacent edges.

Consider the operation $FRUR^{-1}U^{-1}F^{-1}$ and apply it to the starting position. The result is above left. The bottom and middle slices are unchanged. The top corners are all messed up but we are interested in the top edge pieces. That at UL is unmoved. The one that was at UF is now at UR and has been flipped. The one that was at UB is now at UF and has been flipped. The one that was at UR is now at UB but has not been flipped. The purpose of this operation is to flip two of the top edge pieces that were originally at UF and UB. We have

UF_+ UB_+ : $FRUR^{-1}U^{-1}F^{-1}$

where the + suffix means flip. Note that this operation is of order 6.

We can similarly apply $FURU^{-1}R^{-1}F^{-1}$ (this is the reverse operation) with the result as above right. Again we are only interested in the top edge pieces. That at UL is unmoved. The one that was at UF is now at UB and has been flipped. The one that was at UR is now at UF and has been flipped. The one that was at UB is now at UR but has not been flipped. The purpose of this operation is to flip the top edge pieces that were originally at UF and UR. We have

UF_+ UR_+ : $FURU^{-1}R^{-1}F^{-1}$

Note that if all four edges need flipping we can start by applying the first algorithm. This will result in UL and UB needing to be flipped and this can be done by the second algorithm after applying an intermediary U.

So at this stage we should have all four edge pieces of the top layer correctly oriented with the U faces all up. However, it is likely that the edge pieces will not be in the correct places. It will be found that possibly after applying U a few times we have to swap the pairs UF, UR and UB, UL or maybe that UF, UR, UB need to be cycled in that order.

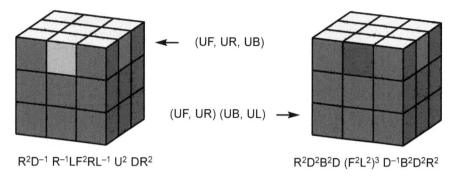

Rearranging the top slice edges.

To cycle (UF, UR, UB) apply $R^2D^{-1}R^{-1}LF^2RL^{-1}U^2DR^2$. This results in the pattern above left. To swap the pairs (UF, UR) and (UB, UL) apply $R^2D^2B^2D$ $(F^2L^2)^3$ $D^{-1}B^2D^2R^2$. This results in the pattern above right.

So in summary we have

(UF, UR) (UB, UL) : $R^2D^2B^2D$ $(F^2L^2)^3$ $D^{-1}B^2D^2R^2$

(UF, UR, UB) : R^2D^{-1} $R^{-1}LF^2RL^{-1}$ U^2 DR^2

We should now have the top slice with all edges correct.

The above moves have the nice property of not messing up the corners. But if the corners are messed up anyway, that is not a great advantage.

The cyclic interchange (UF, UR, UB) is more simply done by $FUF^{-1}UFU^2F^{-1}$. This has 7 rather than 10 moves and only two faces are involved which makes it easier and less likely to go wrong. The result is shown below left.

The interchange (UF, UR) (UB, UL) can be done by two applications of this cyclic interchange with a reorientation of the cube; thus we do $RUR^{-1}URU^2R^{-1}$ followed by $FUF^{-1}UFU^2F^{-1}$. This is 14 moves, the same as for the move which preserves the corners but is perhaps easier to do. This is shown below right.

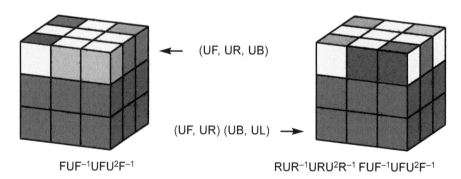

Other moves for the top slice edges.

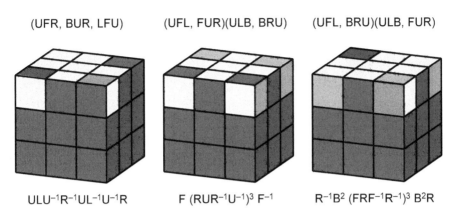

Rearranging the top slice corners.

The next task is to get the corners into place. Various moves are required, much as for the top edges. Three corners might need cycling or two pairs might need interchanging.

To cycle (UFR, BUR, LFU) apply ULU^{-1}R^{-1} UL^{-1}U^{-1}R. This is shown above left. Note that each is twisted one-third of a turn so that the total twist is one. This operation may need to be applied in reverse (take care).

If two pairs need interchanging then this can be done by two applications of the cyclic move or its reverse with appropriate reorientation of the cube. But it is perhaps better to use some other moves as follows.

To interchange the left and right corners in pairs (UFL, FUR) and (ULB, BRU) apply F (RUR^{-1}U^{-1})^3 F^{-1} as shown above centre. Note that this is in fact the operation FRU^{-1}RU^{-1}F^{-1} used to flip the edges UF$_+$ UB$_+$ applied three times. Remember that this operation is of order 6. The flips on the edges cancel out after three applications but the corners are interchanged as desired.

To interchange the diagonally opposite corners we can use the somewhat laborious R^{-1}B^2 (FRF^{-1}R^{-1})^3 B^{2}R shown above right.

So in summary we have

(UFL, FUR) (ULB, BRU) : F (RUR^{-1}U^{-1})^3 F^{-1}	parallel
(UFL, BRU) (ULB, FUR) : R^{-1}B^2 (FRF^{-1}R^{-1})^3 B^{2}R	diagonal
(UFR, BUR, LFU) : ULU^{-1}R^{-1} UL^{-1}U^{-1}R	cycle

Note that we required a special sequence for swapping pairs in parallel as well as across the diagonals. This did not happen in the case of the edges. And we may have to do the cyclic sequence in reverse as well.

Although the top edges and corners will now be in their correct positions, it might be that some of the corner pieces are not oriented correctly and need to be twisted. Two or four corner pieces might need twisting in pairs one clockwise

R^{-1}DRFDF^{-1} U FD^{-1}F^{-1}R^{-1}D^{-1}R U^{-1}

or

(FDF^{-1}D^{-1})^2 U (DFD^{-1}F^{-1})^2 U^{-1}

(RFU)$_+$ (RUB)$_-$

Twisting the top slice corners.

and one anticlockwise. Or three corner pieces may all need twisting the same way. The total twisting required will thus be zero or plus or minus one. Consider

(RFU)$_+$ (RUB)$_-$: R^{-1}DRFDF^{-1} U FD^{-1}F^{-1}R^{-1}D^{-1}R U^{-1}

The + suffix means twist clockwise whereas the − suffix means twist anticlockwise. The result of doing this on a cube at the starting position is shown above. The sequence R^{-1}DRFDF^{-1} does the clockwise twist on RFU. The following U brings RUB to the front and the sequence FD^{-1}F^{-1}R^{-1}D^{-1}R does the anticlockwise twist and the final U^{-1} returns RUB to its proper place.

If we want to do (RFU)$_+$ (LUB)$_-$ that is do the anticlockwise twist on the opposite diagonal corner then replace the U by U^2 and U^{-1} by U^{-2} (that is U^2). So

(RFU)$_+$ (LUB)$_-$: R^{-1}DRFDF^{-1} U^2 FD^{-1}F^{-1}R^{-1}D^{-1}R U^2

It is easy to make mistakes with these moves. Observe that after the first twist the bottom two slices are in a mess; the second twist reverses this mess. Note that the two sequences R^{-1}DRFDF^{-1} and FD^{-1}F^{-1}R^{-1}D^{-1}R are opposites.

Two alternative sequences which can be used instead are (FDF^{-1}D^{-1})^2 for the clockwise twist and its opposite (DFD^{-1}F^{-1})^2 for the anticlockwise twist. Although they have 8 moves each rather than 6, they are perhaps safer. It is a nuisance if a mistake is made at this stage because it will almost inevitably mean starting the whole restoring process from scratch.

If three corners need twisting all the same way then the same technique can be used. Move each corner to FUR in turn and do the appropriate twist. For example

(FUR)$_+$ (BUR)$_+$ (BUL)$_+$: (FDF^{-1}D^{-1})^2 U (FDF^{-1}D^{-1})^2 U (FDF^{-1}D^{-1})^2 U^2

Incidentally we can also use (D^{-1}R^{-1}DR)^2 for the clockwise twist and its reverse (R^{-1}D^{-1}RD)^2 for the anticlockwise twist.

With luck the cube should now be perfectly restored. If not, start again!

Summary of restoration

1. Put white face on top and do edges.

    DR → UR : R²   RD → UR : RUF⁻¹U⁻¹

2. Do white corners.

    FRD → FUR : R⁻¹D⁻¹RD   white right
    DFR → FUR : D⁻¹R⁻¹DR   white front
    RDF → FUR : FD²F⁻¹ D² R⁻¹DR   white down

    Now reverse cube to normal position.

3. Put middle slice edges in place.

    UF → RF : URU⁻¹R⁻¹U⁻¹F⁻¹UF
    UR → FR : U⁻¹F⁻¹UFURU⁻¹R⁻¹

4. Flip top edges so all U faces are up.

    UF₊ UB₊ : FRUR⁻¹U⁻¹F⁻¹   UF₊ UR₊ : FURU⁻¹R⁻¹F⁻¹

5. Rotate U as necessary for the next stage.

6. Put top edges correctly in place.

    (UF, UR) (UB, UL) : RUR⁻¹URU²R⁻¹ FUF⁻¹UFU²F⁻¹
    (UF, UR, UB) : FUF⁻¹UFU²F⁻¹

7. Put top corners correctly in place.

    (UFL, FUR) (ULB, BRU) : F (RUR⁻¹U⁻¹)³ F⁻¹   parallel
    (UFL, BRU) (ULB, FUR) : R⁻¹B² (FRF⁻¹R⁻¹)³ B²R   diagonal
    (UFR, BUR, LFU) : ULU⁻¹R⁻¹ UL⁻¹U⁻¹ R   cycle

8. Twist top corners as necessary.

    (RFU)₊ (RUB)₋ : R⁻¹DRFDF⁻¹ U FD⁻¹F⁻¹R⁻¹D⁻¹R U⁻¹
    (RFU)₊ (LUB)₋ : R⁻¹DRFDF⁻¹ U² FD⁻¹F⁻¹R⁻¹D⁻¹R U²

    Take care over the last stage. You might prefer

    (RFU)₊ (RUB)₋ : (FDF⁻¹D⁻¹)² U (DFD⁻¹F⁻¹)² U⁻¹
    (RFU)₊ (LUB)₋ : (FDF⁻¹D⁻¹)² U² (DFD⁻¹F⁻¹)² U²

The cube group

WE HAVE SEEN that the Slice-squared group has 8 elements and moreover that it is Abelian. All the elements of the Slice-squared group have order 2 and indeed the group is the direct product of three instances of the cyclic group of order 2. So, as discussed in the appendix on Groups, it is $C_2 \times C_2 \times C_2$. The subgroup generated by just R and F has 4 elements and is the four group D_2 which is $C_2 \times C_2$. And of course the subgroup generated by R alone is the cyclic group C_2.

As another example, consider the group generated by the operations R and L. These both have order 4. It is also clear that this group is Abelian since the operations R and L do not interact and indeed this group has 16 elements and is the direct product $C_4 \times C_4$.

The more general Slice group has 768 elements and we noted earlier that 768 factorizes to $2^8 \times 3$. But the whole cube group is enormous and has in fact 43,252,003,274,489,856,000 elements. This factorizes to $2^{27} \times 3^{14} \times 5^3 \times 7^2 \times 11$.

Remember that the order of an element has to be a factor of the order of the group. We have already met operations of various orders. For example

3   $furf = FB^{-1}UD^{-1}RL^{-1}FB^{-1}$

4   R

6   $FRUR^{-1}U^{-1}F^{-1}$

There are also operations of orders such as 5, 7, and 11. This is perhaps very surprising. It is clear that a cube has fourness since a side is a square and that it has threeness since the vertex figure is a triangle. But why fiveness? Maybe something to do with the fact that there are five cubes in a dodecahedron. As for seven and eleven....

It is a good idea at this stage to consider a uniform representation for an operation. Every operation performs one or more permutations on edges and corners. Moreover, there might be flips and twists. We will consider a few of the operations we have met so far.

R^2 => (FR, BR) (UR, DR) (FUR, BDR) (DFR, UBR)

So this comprises two 2-cycles on both edges and corners.

R => (FR, UR, BR, DR) (FUR, UBR, BDR, DFR)

And here we have two 4-cycles. Note carefully that we must match the faces correctly. We do not say that FUR goes to BUR.

Now consider the more elaborate

$FB^{-1}UD^{-1}RL^{-1}FB^{-1}$ => (FR, UF, RU) (FL, UB, RD) (FD, UL, RB) (DL, LB, BD)
FUR_+ BDL_- (DFR, LUF, BRU) (LDF, BLU, DBR)

This is an operation to produce spots on all faces. It changes all edges and corners. The twelve edges perform four 3-cycles. Six of the eight corners form two 3-cycles whereas two of the corners do not move but have a twist, one clockwise and one anticlockwise. Twists of course are also operations of order three and so it is clear why the whole operation has order three.

Now consider

$FRUR^{-1}U^{-1}F^{-1}$ => (FU, UR, UB) (FUR, LUF)$_+$ (BUR, LUB)$_-$

This operation was used to flip the edges of the top layer. The edges form a 3-cycle. The corners perform two 2-cycles. However, they put a twist into the cycles. Thus LUF does not move to FUR but to URF, a clockwise twist. And LUB does not move to BUR but to URB, an anticlockwise twist. So we add a + and − after the cycles to denote these twists. It is clear why this operation has order six. Note that we do this operation three times to interchange the corners of the top layer. Applying it three times means that the edges remain unchanged but the pairs of corners are interchanged and the twists removed.

Consider the operation $FR^2F^{-1}R^2$. This performs a 3-cycle on the edges (FR, BR, UF) and a 5-cycle on the corners (FUR, FRD, BRU, BDR, FLU). Consequently it has order 15. If we do it three times then the edges are not changed and we simply get a 5-cycle on the corners (FUR, BDR, FRD, FLU, BRU). So the cube (the third power) of the operation has order 5 thus

5    $(FR^2F^{-1}R^2)^3$

In a similar way the operation $L^2BRD^{-1}L^{-1}$ performs an 11-cycle on the edges (RF, RU, RB, UB, LD, LB, LU, BD, DF, FL, RD) and a 7-cycle on the corners (FUR, UBR, LDB, LBU, DLF, BDR, DFR). So it leaves one edge and one corner unchanged. This operation thus has order 77. So the 7th power has order 11 and vice versa. So we have found operations of orders 7 and 11.

7    $(L^2BRD^{-1}L^{-1})^{11}$

11   $(L^2BRD^{-1}L^{-1})^7$

For reasons that cannot be explained here, the maximum order of an operation is $1260 = 2^2 \times 3^2 \times 5 \times 7$. A simple example is $RF^2B^{-1}UB^{-1}$. Its description is

(FU, FD, LU, BR, DR, FL, FR)$_+$ (LB, UR)$_+$ (UB, DB)
(DFR, FDL, LUF)$_-$ (URF, BLD, DRB, UBR, BUL)$_+$

Thus it cycles seven of the edges as one group and two pairs as two other groups and then three of the corners as one group and the other five as another group. All of these permutations have twists apart from one. Just one edge is unchanged.

If we draw pictures on the faces of the cube then we discover that for a correct orientation not only do the edges and corners have to be in the right place and oriented correctly but the centres have to be oriented correctly as well. This multiplies the number of operations in the cube group by $2048 = 2^{11}$.

Some cubes have such faces and in order to restore them we first restore the edges and corners in the usual way. We can then simply restore the centres of the faces. Three operations are required for this: one to invert the centre of a single face, one to make quarter turns on adjacent faces, and one to make quarter turns on opposite faces. We have

$(URLU^2R^{-1}L^{-1})^2$ inverts the centre of the U face,

$fur\ F^{-1}\ r^{-1}u^{-1}f^{-1}\ R$ turns R a quarter, and F a negative quarter,

$rf^2r\ U\ rf^2r\ D^{-1}$ turns U a quarter and D a negative quarter.

Remember that r is the quarter turn slice RL^{-1} and so on.

Note that it is not possible to do a quarter turn on one face alone.

Cosets

COSETS are an important feature of group theory and were explained in Appendix E. Remember that a group comprises a set of entities (or elements) with an operation defined upon them. Important features of a group are a) that there is a unit entity which is the starting state I in the case of the cube, and b) that an operation acting on a state produces another state of the group, and c) that every operation has an inverse which when applied gives the state I.

Groups can have subgroups. Thus the full cube group has a subgroup which is the Slice group and that in turn has a subgroup being the Slice-squared group with just 8 elements. Subgroups are themselves truly groups, they include the unit element, I, every element has an inverse, and operating on the elements always produces an element of the subgroup.

Cosets are a bit like subgroups but do not include the unit element I. If we take a subgroup H and an element e not in H then the set e × H is called the left coset and the set H × e is the right coset. If a subgroup is such that the left and right cosets are always the same then the subgroup is said to be a normal subgroup. It so happens that the Slice-squared subgroup is not normal. Nevertheless, in many cases the right and left cosets are the same.

As a first example suppose we start with the cube in the state $(f*r*)^3$ introduced earlier which we can call Z for zigzag. Now suppose we consider the effect of applying the 8 operations of the Slice-squared subgroup which we saw were *I, R, F, U, FU, UR, RF, RFU*. We get the 8 states ZI, ZR, ZF and so on which form the left coset of Z. Note that this set does not include I so it is not a group. But the operations *R, F,* and so on applied to the members of this set always produce a member of the set. The complete left coset is shown opposite. In this case the right coset is identical.

Another interesting coset is obtained by considering the state $(F^2R^2)^3$ which interchanges two edges in the top and bottom slices. We can call it X. The complete left coset is also shown opposite.

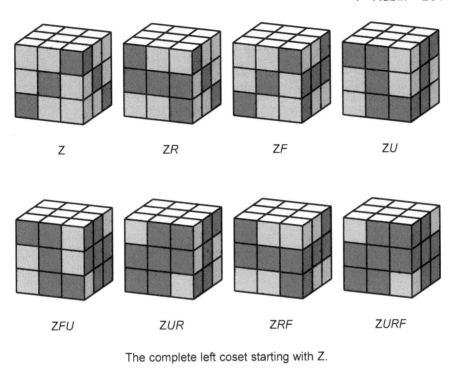

The complete left coset starting with Z.

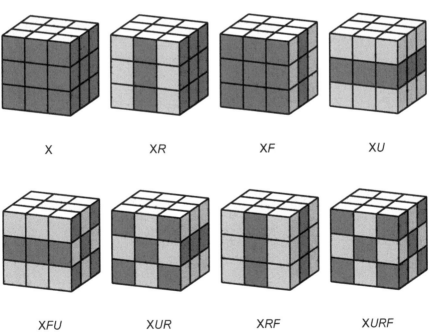

The complete left coset starting with X.

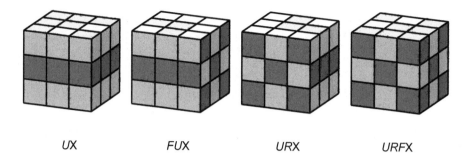

| UX | FUX | URX | URFX |

Those elements of the right coset of X which differ from the left coset.

The right coset is not exactly the same. Thus XU is not the same as UX; the two white edges on the top slice are moved to the other top edges. The other patterns involving U are also different in the same way. These different elements of the right coset are shown above.

It may be recalled from Appendix E on Groups that the left and right cosets may be the same but the elements might be generated in a different order. That does not apply in this case, the cosets are really different.

The reader is invited to make corresponding cosets using other starting positions. In performing manipulations remember that the subgroup consisting of the eight elements I, R, F, U, FU, UR, RF, and RFU is Abelian so that the operations can be performed in any order. Thus ZFU is the same as ZUF and so on.

Finally

HERE ARE SOME other interesting patterns which might be good as a basis for further exploration and demonstration.

The first example shows a pretty hierarchy of nested cubes. The first pattern opposite is the usual start. The second pattern is obtained by $BL^{-1}D^2LDF^{-1}D^2FD^{-1}B^{-1}$ followed by $F^{-1}RU^2R^{-1}U^{-1}BU^2B^{-1}UF$. Note that this second sequence is closely related to the first with B swapping with F^{-1}, L with R^{-1}, and D with U^{-1}. The result is a cube with the top front right 2 by 2 cube being a twisted version of the whole cube.

We can twist the single top front right cube (and its mirror image at the back) to give the third pattern. We can do this by doing B^2 which brings the mirror image up to the top slice and then use the appropriate top corner twists given as item 8 of the restoration process. Take care! Finally we do B again to put the now twisted mirror image back where it belongs.

This is a good arrangement to leave around the house to tempt visitors to meddle with it.

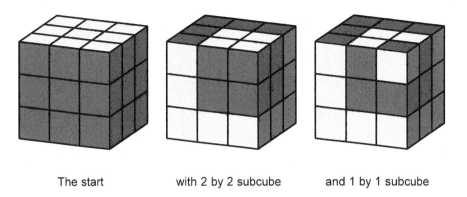

| The start | with 2 by 2 subcube | and 1 by 1 subcube |

The next example shows how we can get large Ts on four faces by (carefully) doing $RLU^2R^{-1}L^{-1}FBU^2F^{-1}B^{-1}U^2$ as shown below. We naturally call this T. And then from T we can produce the complete left Slice-squared coset as shown.

The reader might like to consider whether the right coset is the same. This is somewhat tedious to do from scratch because we have to apply the complex operation used to create T but starting from each Sliced-squared position in turn.

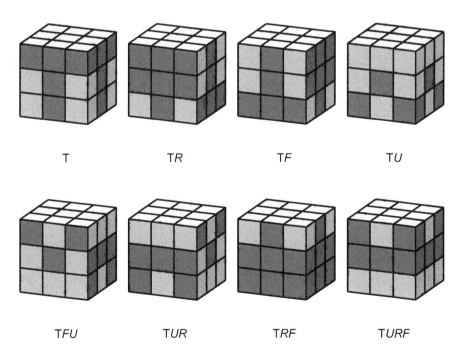

| T | TR | TF | TU |

| TFU | TUR | TRF | TURF |

The complete left coset starting with T.

284 Nice Numbers

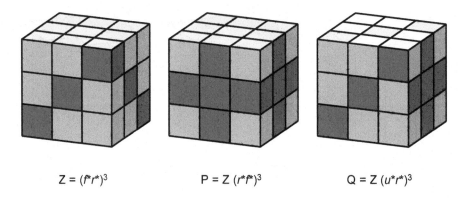

$Z = (f^*r^*)^3$ $\qquad$ $P = Z\,(r^*f^*)^3$ $\qquad$ $Q = Z\,(u^*r^*)^3$

We recall that $Z = (f^*r^*)^3$ produces a zigzag pattern as shown above. If we rotate the cube and then do it again we get further interesting patterns. Note that the rotation of Z about the vertical axis is simply $(r^*f^*)^3$. Thus $(f^*r^*)^3$ followed by $(r^*f^*)^3$ results in plus signs on four faces which we can call P.

However, if we follow Z by $(u^*r^*)^3$ we get plus signs on two faces and zigzags on four; well actually since they all slope the same way perhaps they should be called zigzigs. We can call this Q (for qurious).

P and Q are good starting points for other coset patterns as shown below.

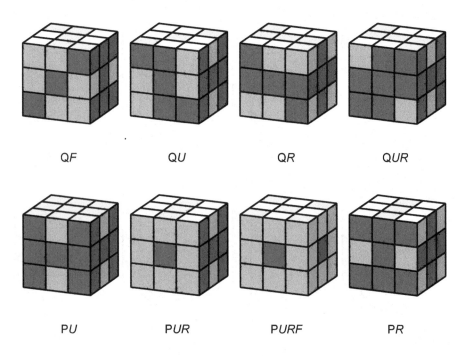

QF $\qquad$ QU $\qquad$ QR $\qquad$ QUR

PU $\qquad$ PUR $\qquad$ PURF $\qquad$ PR

Parts of the left cosets starting with P and Q.

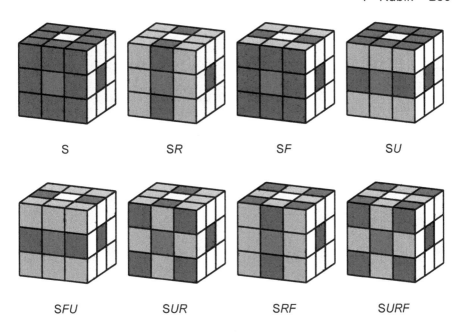

The complete left coset starting with S.

Another good starting point is the pattern of all spots produced by *furf* which we discussed earlier. We naturally call this S. The complete colourful coset is shown above.

Further reading

THERE ARE two highly recommended books on the cube. One is *Rubik's Cubic Compendium* by Rubik and others. The other is *Notes on Rubik's Magic Cube* by David Singmaster.

G Differences

THIS SHORT APPENDIX looks at various so-called difference equations. We encountered such equations when discussing the Tower of Hanoi and Chinese Rings in the final lecture. We also illustrate the technique on the Fibonacci numbers and raise a problem regarding hats.

It should be noted that the term difference equation might seem surprising since explicit differences do not usually appear. Perhaps a better term is recurrence relation.

Rings

IN THE CASE of Chinese Rings, if we require $F(n)$ moves to remove or replace the first n rings as described at the end of the final lecture then we showed that

$$F(n) = F(n-1) + 2F(n-2) + 1$$

This is a quadratic difference equation. We can solve it by using an inspired guess that the solution might be of the form $F(n) = Ax^n + C$. (This is a standard technique or trick.) We get

$$Ax^n + C = Ax^{n-1} + C + 2Ax^{n-2} + 2C + 1$$

We want this to work for any value of n. This means that the terms not involving n must balance. So we deduce

$$C = C + 2C + 1 \quad \text{therefore } C = -1/2.$$

And then, after dividing through by Ax^{n-2}, that leaves the following

$$x^2 = x + 2$$

This is a quadratic equation and its two roots are $x = 2$ and $x = -1$. This means that the general solution has two terms corresponding to Ax^n in the proposed guess and so is

$$F(n) = A2^n + B(-1)^n - 1/2$$

where A and B are two constants which need to be found. In order to do this we consider the two simplest cases. These are that of taking off just 1 ring and that of taking off 2 rings. It takes only 1 move to remove 1 ring and 2 moves to take off 2 rings – remember that ring 2 can be removed at once provided ring 1 is on, so in the case of two rings we take off ring 2 and then ring 1. Putting these values in the general solution gives

$$F(1) = 2A - B - 1/2 = 1 \qquad \text{one ring}$$
$$F(2) = 4A + B - 1/2 = 2 \qquad \text{two rings}$$

These two equations for A and B are easily solved. Adding them together gives

$$6A - 1 = 3$$

so $A = 2/3$ and it then follows that $B = -1/6$. So now we have the final answer that the number of moves required is

$$F(n) = 2/3 \times 2^n - 1/6 \times (-1)^n - 1/2 = 2^{n+1}/3 + (-1)^{n+1}/6 - 1/2$$

It is perhaps neater to get rid of the awkward term $(-1)^{n+1}$ by considering the two cases of odd n and even n separately. This gives

$$F(n) = 1/3(2^{n+1} - 1) \qquad n \text{ odd}$$
$$F(n) = 1/3(2^{n+1} - 2) \qquad n \text{ even}$$

We can check that for $n = 5$ and get $F(5) = 1/3(2^6 - 1) = 1/3 \times 63 = 21$ which agrees with the analysis in the lecture.

Note that if we allow ourselves to move rings 1 and 2 together, then $F(2) = 1$ and we then find that $A = 1/2$ and $B = -1/2$. This gives

$$F(n) = 2^{n-1} \qquad n \text{ odd}$$
$$F(n) = 2^{n-1} - 1 \qquad n \text{ even}$$

which for $n = 5$ becomes 16, and so again agrees with the previous analysis.

Towers

I<small>N THE CASE</small> of the Tower of Hanoi, the calculation is much simpler. To move n discs we first move the top $n–1$ discs to the spare pin then move the one bottom disc and finally move the $n–1$ discs as explained in Lecture 10. As a result we have

$$F(n) = 2F(n-1) + 1$$

Again, suppose the solution is $F(n) = Ax^n + C$ so that

$$Ax^n + C = 2Ax^{n-1} + 2C + 1$$

Matching the terms not involving n gives

$C = 2C + 1$ therefore $C = -1$

And then after dividing through by Ax^{n-1} we simply have

$x = 2$

Note that in this simple example the equation for x is trivially linear and so the formula $F(n) = Ax^n + C$ reduces to

$F(n) = A2^n - 1$

And as before we find the constant A by considering simple cases for which we know the answer. Because the equation is linear we only need the one case that to move a single disc ($n = 1$) only one move is needed. So we get

$F(1) = 2A - 1 = 1$ one disc

and hence A is 1 so that the final answer is that the number of moves required is

$F(n) = 2^n - 1$

which confirms the result stated in Lecture 10.

Fibonacci numbers

AS ANOTHER EXAMPLE consider the Fibonacci numbers once more. Remember that each number in the sequence is the sum of the two previous ones and the first two numbers are both 1. Thus

1, 1, 2, 3, 5, 8, 13, 21, 34, 55, ...

The sequence can be defined by the recurrence relation

$F(n) = F(n-1) + F(n-2)$

As before we assume the solution is of the form $F(n) = Ax^n + C$ where A and C are constants. So we get

$Ax^n + C = Ax^{n-1} + C + Ax^{n-2} + C$

Again the terms not involving n must balance. So we immediately see that C is zero. And then, after dividing the other terms by Ax^{n-2} as in the case of the rings, we have the following

$x^2 = x + 1$ or $x^2 - x - 1 = 0$

This is a quadratic equation and so has two roots. Using the general formula that the roots are $x = (-b \pm \sqrt{(b^2 - 4ac)})/2a$ where in this case a is 1 and b and c are both -1 gives the roots as $x_1 = (1+\sqrt{5})/2$ and $x_2 = (1-\sqrt{5})/2$.

Now $(1+\sqrt{5})/2$ is the golden number (1.618...) often denoted by τ and $(1-\sqrt{5})/2$ is quickly seen to be $1-\tau$ ($-0.618...$) which is also equal to $-1/\tau$. This means that the general solution is

$$F(n) = A\tau^n + B(-1/\tau)^n$$

where A and B are two constants which need to be found. In order to do this we consider the two simplest cases which are that the first two terms are both 1; so $F(1)$ is 1 and $F(2)$ is also 1. Hence we get two equations for A and B thus

$$F(1) = A\tau - B/\tau = 1$$
$$F(2) = A\tau^2 + B/\tau^2 = 1$$

Solving these we find that $A = \tau/(1+\tau^2)$ and $B = -A$. Perhaps surprisingly, A simplifies to just $1/\sqrt{5}$. So we have now found an explicit expression for the nth Fibonacci number namely

$$F(n) = (\tau^n - (-\tau)^{-n})/\sqrt{5}$$

The first term dominates as n gets larger showing that the numbers converge on to powers of the golden number (divided by the square root of five). What is very surprising is that such a gruesome expression riddled with square roots of five (both explicit and implicit in τ) should reduce to a whole number.

Hats

WE NOW LOOK at another interesting puzzle in Dudeney's book *Amusements in Mathematics*. Puzzle number 267 concerns hats. Eight men dine together with an abundance of wine. When they leave the restaurant they are so confused (deranged is an appropriate term) that nobody picks up their own hat. How many different ways could they do this? Dudeney gives the answer as 14833 and two ways of arriving at this number but doesn't explain the reasoning behind them.

Suppose that $F(n)$ is the number of ways in the case of n men. As an example consider the case of five men and five hats. Man A must not have his own hat; there are four ways in which this can be done. Suppose A has B's hat. Then there are two cases to consider. B might or might not have A's hat. Consider these two cases in turn.

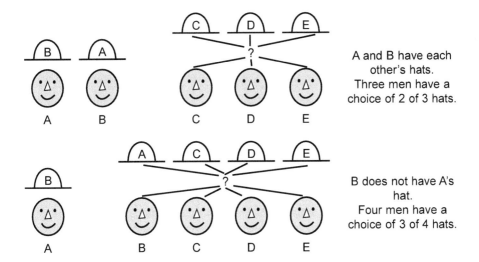

A and B have each other's hats. Three men have a choice of 2 of 3 hats.

B does not have A's hat. Four men have a choice of 3 of 4 hats.

Suppose B has A's hat so that A and B have each other's. The remaining three men C, D, and E therefore must have hats from within that group. Clearly the number of ways that can be done is simply $F(3)$.

On the other hand, suppose B does not have A's hat. Then B has three other hats to choose from (those of C, D, E). Moreover, C, D, E also have three hats to choose from (one is A's but it is sort of proxy for B's). The number of ways in which B, C, D, E can choose hats is clearly $F(4)$. So we end up with

$$F(5) = 4 \times (F(4) + F(3))$$

In the general case it is clear that we have the following recurrence relation

$$F(n) = (n-1) \times ((F(n-1) + F(n-2))$$

If one man dines alone then provided that he remembers to pick up a hat, it cannot be the wrong one so $F(1)$ is zero. If two men dine then the only way to pick up wrong hats is to take each other's hat and so $F(2)$ is one. We can now easily compute successive values and obtain the answer for eight men. Thus

$F(3) = 2 \times (0 + 1) = 2$
$F(4) = 3 \times (1 + 2) = 9$
$F(5) = 4 \times (2 + 9) = 44$
$F(6) = 5 \times (9 + 44) = 265$
$F(7) = 6 \times (44 + 265) = 1854$
$F(8) = 7 \times (265 + 1854) = 14833$

The other recurrence relation for the problem also mentioned by Dudeney is

$$F(n) = n \times F(n-1) + (-1)^n$$

n	1	2	3	4	5	6	7	8
!n	0	1	2	9	44	265	1854	14833
$n!$	1	2	6	24	120	720	5040	40320
!$n/n!$	0	0.5	0.3333	0.3750	0.3666	0.3680	0.3678	0.3678

Derangement!

and it is easy to see that it produces the same sequence of values.

Finding a general formula is not simple. Unlike the other difference equations, the coefficients are not constant but depend upon n.

This problem is an example of a derangement in which everything is in the wrong place. It was first formulated and solved by the French mathematician Pierre de Montmort (1678–1719). It is also discussed by Rouse Ball in his *Mathematical Recreations and Essays* in terms of putting several letters all in the wrong envelopes. Imagine the fuss if they were letters to several mistresses!

The solution is given by a function often known as the subfactorial function and denoted by an exclamation mark in front of the argument. The value of the subfactorial function is

$$!n = n! \times (1 - 1/1! + 1/2! - 1/3! + \cdots \pm 1/n!)$$

We can try it for say $n = 4$ and get

$$!4 = 24 \times (1 - 1 + 1/2 - 1/6 + 1/24) = 24 - 24 + 12 - 4 + 1 = 9$$

It is not too difficult but a bit tiresome to show that it satisfies the two difference equations. We leave this to the reader.

The formula can perhaps more amusingly be written as

$$\frac{!n}{n!} = 1 - \frac{1}{1!} + \frac{1}{2!} - \frac{1}{3!} + \cdots \pm \frac{1}{n!}$$

The finite series is in fact the first few terms of the infinite series for $1/e$ where e is the base of natural logarithms as mentioned in the lecture on Primes. Remember that the series for e^x is

$$e^x = 1 + \frac{x}{1!} + \frac{x^2}{2!} + \frac{x^3}{3!} + \cdots$$

so putting $x = -1$ gives the series above. Note that $1/e$ is 0.36787944....

As a consequence, as n gets large we find that !n is closely approximated by $n!/e$. The table above shows values for !n and $n!$ and their ratio.

So the probability of everybody having the wrong hat is about 37%. An interesting consequence of the convergence is that even with a large number of items the probability remains much the same.

Suppose we take two packs of 52 cards arranged in order and shuffle just one pack. Then deal the cards out in parallel into two piles. The chance of at least one pair matching is 63% which seems surprisingly large.

Further reading

AN EXCELLENT DISCUSSION of the Fibonacci numbers will be found in *An Introduction to Geometry* by Coxeter. He states that the explicit formula was first discovered by J P M Binet in 1843. However, according to Knuth in *The Art of Computer Programming, Vol 1*, it was actually found by the Huguenot mathematician Abraham de Moivre (1667–1754) in 1718.

Various websites such as Wikipedia contain much information on difference equations.

H Chinese Remainders

IN THIS APPENDIX we look at the so-called Chinese Remainder Theorem which is an intriguing backwater of number theory. The name arose because of its use in solving several historical puzzles in early Chinese literature concerning remainders. A typical puzzle is

> A basket contains a number of eggs. If they are taken out two at a time then just one egg is left. If they are taken out three at a time then two eggs are left. Similarly, if they are taken out four, five, or six at a time then three, four, and five are left respectively. However, if they are taken out seven at a time, then no eggs are left. How many eggs were in the basket? Strictly, what is the smallest number of eggs that could have been in the basket?

So the number of eggs has to satisfy several linear congruences at the same time. We first explore how to solve single linear congruences and then look at the question of simultaneous congruences.

Linear congruences

IN THE FINAL LECTURE when discussing the RSA algorithm we introduced Diophantine equations which are equations where we require that the unknown values be integers. In particular we noted that the equation

$$ax + by = c$$

where a, b, and c are integers has solutions where x and y are integers only if the gcd of a and b is a factor of c. We then showed how the Euclidean algorithm could be used to find the solutions.

If we have a solution then it is clear that

$$ax \equiv c \bmod b$$

since this simply means that ax and c differ by a multiple of b and indeed they differ by y times b.

So the Euclidean algorithm can be used to solve equations of the form

$$ax \equiv b \bmod n$$

provided that the gcd of a and n is a factor of b.

As a trivial example suppose we wish to solve

$$3x \equiv 1 \bmod 7$$

The gcd of 3 and 7 is 1 so there is a solution. The Euclidean algorithm process on 3 and 7 is trivial. We have

$7 = 2 \times 3 + 1$ line 1
$3 = 3 \times 1 + 0$ the gcd is 1 because the remainder is 0

So unwinding

$1 = 7 - 2 \times 3$ from line 1

and rearranging

$$3 \times (-2) \equiv 1 \bmod 7$$

So $x = -2$ is a solution but because we are working mod 7 it follows that $x = 5$ is also a solution. It is clearly best to write

$$x \equiv 5 \bmod 7$$

as the overall solution.

As another example consider

$$5x \equiv 7 \bmod 17$$

The gcd of 5 and 17 is 1 which is a factor of 7 so again there is a solution. The Euclidean algorithm process is

$17 = 3 \times 5 + 2$ line 1
$5 = 2 \times 2 + 1$ line 2
$2 = 2 \times 1 + 0$ the gcd is 1 because the remainder is 0

So unwinding

$1 = 5 - 2 \times 2$ from line 2
$ = 5 - 2 \times (17 - 3 \times 5)$ from line 1
$ = 7 \times 5 - 2 \times 17$ gathering 5s

and rearranging

$5 \times 7 \equiv 1 \bmod 17$ now multiply by 7 to get
$5 \times 49 \equiv 7 \bmod 17$

So $x = 49$ is a solution and we can now reduce by 34 to get

$$x \equiv 15 \bmod 17$$

That was tedious, so better check it. And indeed $5 \times 15 = 75 = 7 + (4 \times 17)$. So we have seen how to reduce any equation of the form $ax \equiv b$ mod n into one of the form $x \equiv d$ mod n using the Euclidean algorithm.

Note that if n is small it is often quicker (and possibly more reliable!) to take a sort of iterative approach. For example, if we start with $5x \equiv 7$ mod 17 we can consider various values of x starting with $x = 1$ as follows

$5 \times 1 \equiv 5$ mod 17
$5 \times 2 \equiv 10$ mod 17
$5 \times 3 \equiv 15$ mod 17
$5 \times 4 \equiv 3$ mod 17

What we are trying to do is find a value of x so that the right hand side is 7 mod 17. Clearly 24 mod 17 will do. So we just multiply 5×4 by 8 and get

$5 \times 32 \equiv 24$ mod $17 \equiv 7$ mod 17

So 32 is a possible answer. We reduce it modulo 17 as usual and get $x = 15$ as before.

Simultaneous equations

WE NOW CONSIDER how to find a solution to several simultaneous linear congruences. That is a group of equations of the form

$x \equiv a_1$ mod n_1; $\quad x \equiv a_2$ mod n_2; $\quad x \equiv a_3$ mod n_3; $\quad ...$

We assume that the moduli n_1, n_2, n_3, ... are all relatively prime, that is no two have any factors in common. For example they might be 3, 4, and 5.

We form the number N which is the product of all the n_i. In the case of 3, 4, and 5, N would be 60. The Chinese Remainder Theorem says that such a set of equations always has a solution and it is unique modulo N. We illustrate the truth of the theorem by showing the construction of the solution.

We first form the numbers

$N_1 = N/n_1$; $\quad N_2 = N/n_2$; $\quad N_3 = N/n_3$; $\quad ...$

In the case of 3, 4, and 5 they are

$N_1 = 60/3 = 20$; $\quad N_2 = 60/4 = 15$; $\quad N_3 = 60/5 = 12$

Now since the numbers n_i are all relatively prime to each other it follows that the gcd of N_i and n_i is always 1. Thus in the example we have gcd(20, 3) = gcd(15, 4) = gcd(12, 5) = 1.

We now find the solutions to the individual equations

$$N_i x_i \equiv 1 \bmod n_i$$

which we know exist because the gcd of N_i and n_i are all 1. In the case of the example this means solving the three equations

$$20x_1 \equiv 1 \bmod 3; \quad 15x_2 \equiv 1 \bmod 4; \quad 12x_3 \equiv 1 \bmod 5$$

which we can do by Euclid or by "inspection" as described in the previous section. The answers are

$$x_1 \equiv 2 \bmod 3; \quad x_2 \equiv 3 \bmod 4; \quad x_3 \equiv 3 \bmod 5$$

We now form the sum of all the $a_i N_i x_i$, that is $\Sigma\, a_i N_i x_i$

$$x = a_1 N_1 x_1 + a_2 N_2 x_2 + a_3 N_3 x_3 + \cdots$$

and, perhaps amazingly, that is the answer since it satisfies all the original equations. To see this, consider the first term and the first equation. We have

$$x \bmod n_1 \equiv a_1 N_1 x_1 \bmod n_1 + a_2 N_2 x_2 \bmod n_1 + a_3 N_3 x_3 \bmod n_1 + \cdots$$

Now x_1 was constructed so that $N_1 x_1 \equiv 1 \bmod n_1$, so the first term is simply a_1 mod n_1. However, all the other terms are zero because $N_2, N_3, \ldots$ are all divisible by n_1 because of their construction; remember that N is the product of all the n_i and N_2 is simply N/n_2 and hence is divisible by n_1. In other words $N_2 \bmod n_1$ is zero. So we end up with

$$x \equiv a_1 \bmod n_1$$

and similarly for the other original simultaneous equations.

To conclude the example suppose that a_1, a_2, and a_3 are 1, 2, and 4, respectively, so that the equations are

$$x \equiv 1 \bmod 3; \quad x \equiv 2 \bmod 4; \quad x \equiv 4 \bmod 5$$

Now we have already seen that x_1, x_2, and x_3 are 2, 3, and 3, respectively. So we can put it all together and obtain

$$x = 1 \times 20 \times 2 + 2 \times 15 \times 3 + 4 \times 12 \times 3 = 40 + 90 + 144 = 274$$

We can then reduce x by multiples of 60 to obtain the general solution

$$x \equiv 34 \bmod 60$$

Finally, we check that this is correct by noting that

$$34 = 11 \times 3 + 1; \quad 34 = 8 \times 4 + 2; \quad 34 = 6 \times 5 + 4$$

Pirates

AS A FIRST EXAMPLE consider the following ancient Chinese problem. A band of 17 pirates stole a sack of gold coins. They tried to divide this loot into equal portions but found that there were 3 coins over. They squabbled and one was killed. Now with 16 pirates they tried again to divide it into equal portions but found there were 10 coins over. Again they squabbled and another pirate was killed. However, they now found that the loot could be divided equally. What was the least number of coins in the sack?

So we have to solve the three equations

$$x \equiv 3 \bmod 17; \quad x \equiv 10 \bmod 16; \quad x \equiv 0 \bmod 15$$

We first note that 17, 16, and 15 are mutually coprime so that the Chinese Remainder Theorem can be applied directly.

So using the general notation we have

$N = 17 \times 16 \times 15 = 4080$
$N_1 = 4080/17 = 240; \quad N_2 = 4080/16 = 255; \quad N_3 = 4080/15 = 272$

Next we have to solve the three equations

$$240x_1 \equiv 1 \bmod 17; \quad 255x_2 \equiv 1 \bmod 16; \quad 272x_3 \equiv 1 \bmod 15$$

After perhaps a bit of a struggle (see below) we find that the solutions are

$x_1 = 9$ $9 \times 240 = 2160 = 127 \times 17 + 1$
$x_2 = 15$ $15 \times 255 = 3825 = 239 \times 16 + 1$
$x_3 = 8$ $8 \times 272 = 2176 = 145 \times 15 + 1$

So the raw answer is

$$x = 3 \times 240 \times 9 + 10 \times 255 \times 15 + 0 \times 272 \times 8 = 6480 + 38250 = 44730$$

Finally, we reduce this mod 4080 and find that the smallest possible number of coins is 3930.

$240 \times 1 \equiv 2 \bmod 17$ $240 \times 9 \equiv 18 \bmod 17$ which is $1 \bmod 17$ so the answer is 9.	$255 \times 1 \equiv 15 \bmod 16$ $255 \times 1 \equiv -1 \bmod 16$ $255 \times 15 \equiv -15 \bmod 16$ which is $1 \bmod 16$ so the answer is 15.	$272 \times 1 \equiv 2 \bmod 15$ $272 \times 8 \equiv 16 \bmod 15$ which is $1 \bmod 15$ so the answer is 8.

Solving the three equations by inspection.

Eggs

AND NOW we can tackle the egg problem introduced at the start of this appendix. Remember that if the eggs are taken out of the basket 2, 3, 4, 5, or 6 at a time then 1, 2, 3, 4, or 5 remain, but if they are taken out 7 at a time then none remain. So we have to solve the six simultaneous equations as follows

$n \equiv 1 \bmod 2$
$n \equiv 2 \bmod 3$
$n \equiv 3 \bmod 4$
$n \equiv 4 \bmod 5$
$n \equiv 5 \bmod 6$
$n \equiv 0 \bmod 7$

The first thing to note is that the moduli are not mutually coprime and so we cannot use the Chinese Remainder Theorem straight away. The problem lies with the following

$n \equiv 1 \bmod 2$
$n \equiv 2 \bmod 3$
$n \equiv 3 \bmod 4$
$n \equiv 5 \bmod 6$

The lcm (least common multiple) of 2, 3, 4, and 6 is 12, so consider values mod 12. We find that

1, 3, 5, 7, 9, 11	satisfy $n \equiv 1 \bmod 2$
2, 5, 8, 11	satisfy $n \equiv 2 \bmod 3$
3, 7, 11	satisfy $n \equiv 3 \bmod 4$
5, 11	satisfy $n \equiv 5 \bmod 6$

Only 11 satisfies all four equations so they can be replaced by the one equation $n \equiv 11 \bmod 12$. In other words we now have to solve just the three equations

$n \equiv 4 \bmod 5$
$n \equiv 11 \bmod 12$
$n \equiv 0 \bmod 7$

The moduli 5, 12, and 7 are now mutually coprime so we can use the Chinese Remainder Theorem. Using the standard notation we have

$a_1 = 4$; $a_2 = 11$; $a_3 = 0$; $N = 5 \times 12 \times 7 = 420$
$N_1 = 84$; $N_2 = 35$; $N_3 = 60$

We now have to solve

$84x_1 \equiv 1 \bmod 5$; $35x_2 \equiv 1 \bmod 12$; $60x_3 \equiv 1 \bmod 7$

The answers are easily found to be

$$x_1 = 4; \quad x_2 = 11; \quad x_3 = 2$$

from which we get (using . for multiply)

$$n = \Sigma\, a_i N_i x_i = 4.84.4 + 11.35.11 + 0.60.2 = 1344 + 4235 = 5579$$

Finally, we reduce this modulo 420 and obtain 5579 − 13×420 = 119. So the smallest possible number of eggs in the basket is 119.

A neat shortcut is to note that the lcm of 2, 3, 4, 5, 6 minus 1 must solve all the equations except the last since we are one egg short in all those cases. So 59 (or indeed −1) satisfies all equations except the last. Now 59 is 3 mod 7 and 60 is 4 mod 7, so 59+60 = 119 is 0 mod 7 and therefore satisfies all the equations. So once more we find that 119 is the answer.

Squares

As a final example, consider the problem of finding three consecutive positive integers each of which has a square prime factor. Clearly the squares must be different and mutually coprime otherwise the numbers could not be consecutive. Ignore the trivial solution of 7, 8, 9 where 7 is a multiple of 1, 8 is a multiple of 4, and 9 is a multiple of 9. Also, 1 is not considered to be a prime number anyway.

Suppose the integers are $x–2$, $x–1$, x (this makes for much easier working than x, $x+1$, $x+2$ because the numbers involved are smaller).

The obvious choice is using the squares of 2, 3, 5. They can be arranged in six ways, namely

$$2, 3, 5; \quad 2, 5, 3; \quad 3, 2, 5; \quad 3, 5, 2; \quad 5, 2, 3; \quad 5, 3, 2$$

Taking 2, 3, 5, we have

$$x \equiv 2 \bmod 4; \quad x \equiv 1 \bmod 9; \quad x \equiv 0 \bmod 25$$
$$N = 4.9.25 = 900$$
$$N_1 = 225; \quad N_2 = 100; \quad N_3 = 36; \quad \text{so we have}$$
$$225x_1 \equiv 1 \bmod 4; \quad 100x_2 \equiv 1 \bmod 9; \quad 36x_3 \equiv 1 \bmod 25$$

giving $x_1 = 1$; $x_2 = 1$; $x_3 = 16$. So finally $x = \Sigma\, a_i N_i x_i =$

$$2.225.1 + 1.100.1 + 0.36.16 = 550$$

So the numbers are 548 (= 137×4), 549 (= 61×9), 550 (= 22×25).

It is easy to do the other combinations of 2, 3, 5 because much of the calculation is unchanged. The results are

2, 3, 5	548, 549, 550
2, 5, 3	124, 125, 126
3, 2, 5	423, 424, 425
3, 5, 2	774, 775, 776
5, 2, 3	475, 476, 477
5, 3, 2	350, 351, 352

This is possibly good fun for a wet Sunday afternoon. A next step might be

3, 5, 7	7299, 7300, 7301

Why not try some nice prime numbers such as 37, 41, 43?

J Mersenne

THIS FINAL APPENDIX looks in detail at some of the various keyboards described by Mersenne leading up to the 31-note keyboard mentioned in the lecture on Music. In order to gain a good understanding of the relationships between various scales we introduce a Tonnetz schema for describing the notes provided.

The story concludes by looking at the equitempered version of Tonnetz which reveals that much music only works because 12 is divisible by 3 and 4.

Mersenne's 31-note keyboard

IN HIS HUGE WORK, *Harmonie Universelle*, Mersenne described a number of keyboards, ending up with one of 31 keys per octave. The simpler keyboards have 12, 16, 18, and 26 keys in the octave. These counts do not include the final note and so sometimes he refers to them as having 19 or 27 keys and so on. Recall that the traditional keyboard has 12 keys per octave.

It so happens that the frequencies of the notes in the simpler keyboards are all in the 31-note keyboard – in other words the keyboards are all subsets of the most elaborate one. In this description we will show the keyboards in a style which enables them to be related easily.

Mersenne did not use frequencies such as 240 and 400 for his description but instead what are in effect wavelengths. Thus in the 31-note keyboard C (240) is marked 144000 and A (400) is marked 86400. Observe that 240×144000 and 400×86400 are both equal to 34560000.

Maybe he did this because he was familiar with organs where the longest pipes have the lowest notes (smallest frequencies). A frequency of 240 has a wavelength of about 4 feet so maybe we can think of 144000 as being the length in thousandths of centimetres. He probably chose such large numbers in order to avoid fractions.

We start by showing the full 31-note keyboard so that we can then relate the simpler keyboards to it. It was laid out as shown overleaf. The white keys for D to B are duplicated in two rows. The normal scale of C major uses the upper ones. The lower ones are a syntonic comma lower. Thus there are 13 white keys (not counting C twice). There are 18 black keys again more or less arranged as two rows. There are 5 between C and E and 13 between E and C'. The black keys in the upper row are generally a syntonic comma higher than a corresponding key in the lower row.

The keys are numbered in increasing order of frequency or decreasing order of wavelength. Thus we start with C_1 and end with C_{32}. The two versions of D are marked D_4 and D_5. The black keys are marked with an X, so that between C_1

304 Nice Numbers

	135000 X_3		120000 X_8		102400 X_{15}	100000 X_{17}		90000 X_{22}	81920 X_{26}	80000 X_{28}	
	138240 X_2	122880 X_6	121500 X_7	110592 X_{11}	103680 X_{14}	101250 X_{16}	92160 X_{20}	91125 X_{21}	82944 X_{25}	81000 X_{27}	73728 X_{31}
		128000 D_5	115200 E_{10}		108000 F_{13}		96000 G_{19}		86400 A_{24}	76800 B_{30}	
144000 C_1		129600 D_4	116640 E_9		109350 F_{12}		97200 G_{18}		87480 A_{23}	77760 B_{29}	72000 C_{32}

Mersenne's 31-note keyboard with his wavelength numbering.

and D_4 we have X_2 and X_3. Incidentally, Mersenne marked X_{27} and X_{28} shown above as B_{27} and B_{28} and what we are calling B_{29} and B_{30} he marked with the natural sign. The reasons for this need not bother us.

The keyboard above is marked with Mersenne's numbers. There seem to be a couple of printer's errors in the original which are corrected here. Mersenne gives D_5 as 12800, obviously just a zero missing but more seriously he gives A_{23} as 87930. Now a syntonic comma below A_{24} is 87480 and since all the other keys in the lower white row are a syntonic comma below the corresponding one in the upper row presumably somewhere the figures 48 were transcribed as 93 which is understandable if one looks at the style of the original text.

The diagram below shows the keyboard with the more familiar frequencies and again A_{23} has been corrected.

	256 X_3		288 X_8		$337\frac{1}{2}$ X_{15}	$345\frac{3}{5}$ X_{17}		384 X_{22}	$421\frac{7}{8}$ X_{26}	432 X_{28}	
	250 X_2	$281\frac{1}{4}$ X_6	$284\frac{4}{9}$ X_7	$312\frac{1}{2}$ X_{11}	$333\frac{1}{3}$ X_{14}	$341\frac{1}{3}$ X_{16}	375 X_{20}	$379\frac{7}{27}$ X_{21}	$416\frac{2}{3}$ X_{25}	$426\frac{2}{3}$ X_{27}	$468\frac{3}{4}$ X_{31}
		270 D_5	300 E_{10}		320 F_{13}		360 G_{19}		400 A_{24}	450 B_{30}	
240 C_1		$266\frac{2}{3}$ D_4	$296\frac{8}{27}$ E_9		$316\frac{4}{81}$ F_{12}		$355\frac{5}{9}$ G_{18}		$395\frac{5}{81}$ A_{23}	$444\frac{4}{9}$ B_{29}	480 C_{32}

Mersenne's 31-note keyboard with our frequencies.

When we look at the simpler keyboards we will use the letters and suffixes as on the 31-note keyboard in order to ease comparison. However, when we manage to identify say X_2 as C# then we will write it as C#$_2$.

In the lecture on Music we observed that in order to provide just major scales from 6 flats to 6 sharps we needed 31 notes per octave. However, Mersenne's keyboard does not provide the expected notes at all. Hopefully, the reason why will become clear.

A schema

IT IS CONVENIENT to introduce a geometrical schema for understanding the relationships between notes. There are six important intervals surrounding any note. First there is the all important fifth (3/2) and its complement, the fourth (4/3). A fifth plus a fourth make an octave of course so that going up a fifth is the equivalent of going down a fourth. We can lay out related notes thus

$$B\flat — F — C — G — D — A$$

Going to the right goes up a fifth whereas going to the left goes up a fourth, due allowance being made for shifting by an octave where necessary.

A fifth can be subdivided into a major third (5/4) and a minor third (6/5). And the complements of these are the minor sixth (8/5) and major sixth (5/3), respectively. We can depict these other intervals by extending the diagram into two dimensions. Going up a major third is depicted by a shift in a NE direction whereas a minor third is depicted by a shift SE. Similarly, NW depicts a major sixth and SW depicts a minor sixth. The pattern around C thus becomes

```
        A   E
         \ /
    F — C — G
         / \
        A♭  E♭
```

This pattern can be extended indefinitely and we end up with the diagram overleaf which shows the frequencies that we have been using. Note that they have been standardized so that they all lie in the octave from 240 to 480. The rows of the schema are essentially all the same. By going down one row we find the same notes shifted three and a half places to the right. However, the frequencies change. Those in the lower row are exactly a syntonic comma higher than those in the row above. So the row below that with C = 240 is shifted to the right and has C = 240 × 81/80 = 243.

Going up two rows raises all notes by a sharp (a ratio of 25/24 – a small semitone) whereas going down two rows lowers the notes by a flat.

306 Nice Numbers

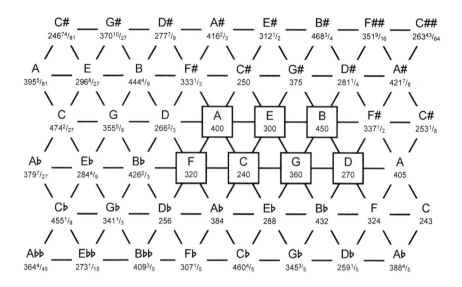

This schema is an example of a Tonnetz lattice which is derived from a description originally conceived by Euler in 1739. (Tonnetz is German for tone-network.) Once again we see the close interrelation between mathematics and music and how it has been explored by eminent mathematicians. Incidentally, the plural of schema is schemata (just as the plural of lemma is lemmata).

Much can be observed from the schema. Any triangle of three notes forms a chord comprising a third and a fifth and known as a triad. If the base is down as for example CEG then it is a major triad. But if the base is up as in the case of CE♭G then it is a minor triad.

Any cluster of seven notes comprising four notes in a row and the three in the row above forms the seven notes needed for a just major scale. The notes thus form the pattern

$$\begin{matrix} & 6 & 3 & 7 & \\ 4 & 1 & 5 & 2 & \end{matrix}$$

where the numbers indicate the order of the notes within the scale.

The notes for C major that we have been using are outlined in the schema above. Note that the schema shows three instances of C with frequencies $474^2/_{27}$, 240, and 243. Moreover, the D in the third row is D_4 ($266^2/_3$) in Mersenne's keyboard, whereas the D in the fourth row is D_5 (270).

The just scale of C comprises the four notes FCGD from the fourth row and AEB from the third row. Adding a sharp moves the pattern one step to the right, whereas adding a flat moves the pattern one step to the left. So this arrangement only uses two rows of the schema and in order to cover from 5 flats to 5 sharps has to be extended as shown above opposite. It requires 27 keys per octave.

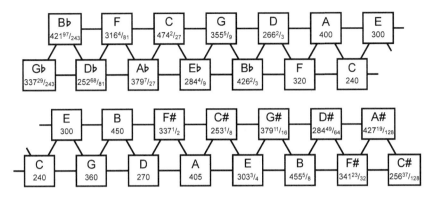

Two rows of the schema covering all just major scales from 5 flats to 5 sharps.

The Pythagorean scale of C.

The Pythagorean scale on the other hand is very simple and starting from C at 240 all the notes are in one row. Going up by a sharp moves the row one place to the right and going down a flat moves it one place to the left. The case of C is also illustrated above. Note that the A, E, B which were on the row above in the just scale are moved down and along for the Pythagorean scale; this raises them by a syntonic comma.

The minor just scales add another twist. We have been describing the minor scales using C minor with three flats. The proper companion to C major is A minor. The required notes for C major and A minor are shown below with red for major and blue for minor and dashed for both. This needs three rows of the schema and moreover D is duplicated. Many more keys are required.

We will now explore some of the various keyboards devised by Mersenne using the schema and so hopefully understand the rationale for the keyboards.

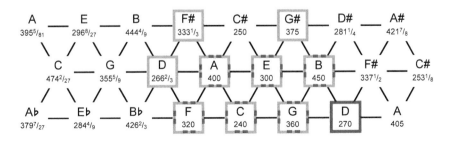

The pattern of keys for C major and A minor combined.

The 12-note keyboards

THE FIRST 12-NOTE KEYBOARD is on page 350 of *Harmonie Universelle*, vol III and is depicted below together with the schema showing the used notes outlined. The keys have been left in the same position as on the 31-note keyboard but of course they are in the traditional place in Mersenne's diagram.

The white notes provide the just scale of C major and form the central pattern in the schema. Additional harmonious major triads are then provided by

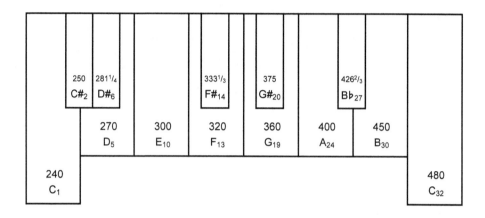

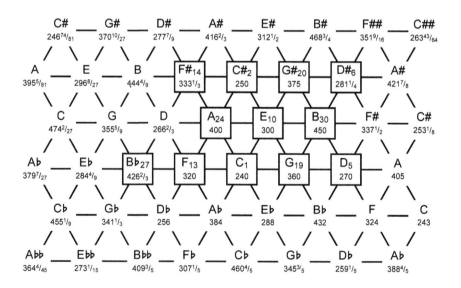

A 12-note keyboard and schema.

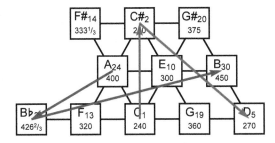 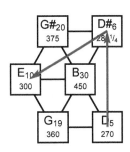

A major tone CD divided into a small semitone and a Great Limma (blue) and a major tone AB divided into a diatonic semitone and a Larger Limma (red).

A minor tone DE divided into a small semitone and a diatonic semitone.

C# and G# in row 2 of the schema. And then F# and D# provide harmonious minor triads. The reason for the choice of B♭ is less clear; the just scale with one flat cannot be done because it requires the D in row three.

No major scales other than C major are possible (this is easily seen by considering the necessary pattern of four notes in one row with three in the row above). And in the case of minor scales the only possibility is the descending melodic form of C# minor. It is probably the case that Mersenne's objective was to provide harmonious chords rather than a range of just scales. Indeed, all possible major triads based on two natural notes are provided.

Mersenne was very enthusiastic regarding the subdivision of the various intervals. Indeed most of his description is about the intervals. The intervals CD and FG (major tones of 9/8) are both split 25/24 (small semitone) and 27/25 (Great Limma) by C# and F# respectively. The intervals DE and GA (minor tones of 10/9) are both split into 25/24 and 16/15 (diatonic semitone) by D# and G# respectively. The other major tone AB is split into 16/15 and 135/128 (Larger Limma) by B♭.

The reader might find it interesting to trace these small intervals on the schema (but if not please move on). A major tone (such as $C_1 D_5$) is two notes to the right, whereas a minor tone (such as $D_5 E_{10}$) is one note NW and one note left. See the diagrams above.

The small semitone (25/24 = 71 cents, such as $C_1 C\#_2$) is two rows up, the Larger Limma (135/128 = 92 cents, such as $B\flat_{27} B_{30}$) is four notes to the right and one NE, the diatonic semitone (16/15 = 112 cents, such as $A_{24} B\flat_{27}$) is one note SW and one note left, and the Great Limma (27/25 = 133 cents, such as $C\#_2 D_5$) is two notes SE and one note right.

Mersenne also describes a second 12-note keyboard which is slightly different and then a 16-note keyboard which in essence is the merger of the two 12-note keyboards. The rationale for these keyboards is not at all clear so we shall move on.

An 18-note keyboard

THE NEXT KEYBOARD is promisingly symmetric with two central rows of five notes and rows of four notes above and below. D is now duplicated and so are F# and B♭. Neither A# nor G♭ are provided. This makes some sort of sense. A# and G♭ are the extreme sharp and flat keys, whereas F# and B♭ are the first to arise.

In summary, this keyboard allows five major scales and four minor scales. It also provides all the major and minor triads based on two natural notes as well as all triads including just one natural note such as $B_{30}F\#_{15}D\#_6$.

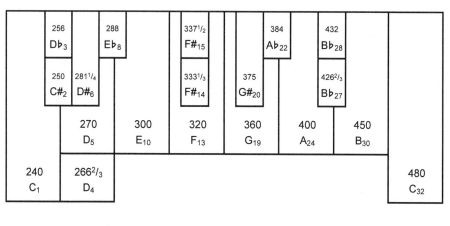

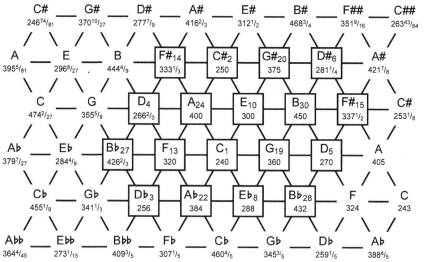

The schema for the 18-note keyboard shows a symmetrical pattern on four rows.

A 26-note keyboard

THE NEXT KEYBOARD adds eight further notes. It duplicates G as well as D and adds many more black keys. The schema now has two central rows of six keys flanked by rows of five keys plus a top row of three keys and a bottom row of just one key. The pattern is not at all symmetric.

The following major scales are permitted: C, F (1 flat), B♭ (2 flats), A♭ (4 flats), D♭ (5 flats), A (3 sharps), E (4 sharps), C# (7 sharps), and the following full minor scales are permitted: A, C (3 flats), F (4 flats), B♭ (5 flats), E (1 sharp), C# (4 sharps) and various other parts of minor scales are possible.

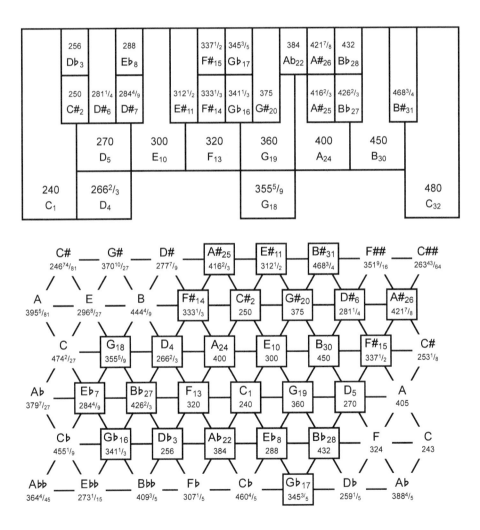

The 26-note keyboard and schema.

The 31-note keyboard

AND FINALLY we come to the 31-note keyboard. This completes the additional white keys a syntonic comma below the others by adding E_9, F_{12}, A_{23}, and B_{29}. It also adds another A♭ (X_{21}). In almost all cases the doubled black notes are simply two versions of the same note but a syntonic comma different.

The resulting schema is shown below. The first observation to make is that the pattern has a hole! The C that is a syntonic comma below the existing C is not included and yet all notes around it are. Why did Mersenne do this? Maybe because it seemed wrong to have two ways to play C major.

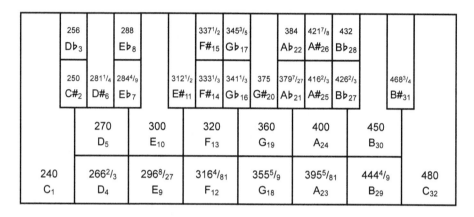

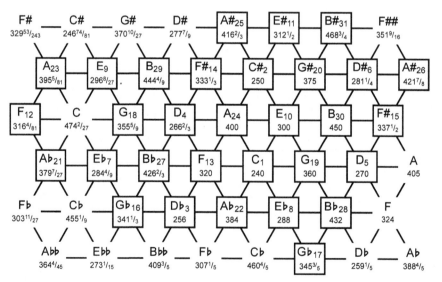

The 31-note keyboard and schema.

The following major scales are possible: C, F (1 flat), B♭ (2 flats), A♭ (4 flats), D♭ (5 flats), D (2 sharps), A (3 sharps), E (4 sharps), C# (7 sharps).

So if the goal were to provide all major keys then there are nasty gaps. Moreover, the scales just mentioned do not use A_{23}, E_9, $A\#_{26}$, F_{12}, $A\flat_{21}$, or $G\flat_{17}$. Many of these notes are useless because of the missing C.

If we consider the minor scales then the following can be done in full: A, D (1 flat), C (3 flats), F (4 flats), B♭ (5 flats), E (1 sharp), C# (4 sharps).

Intervals

MERSENNE described his keyboards largely in terms of the small intervals between adjacent notes. We have already described some of these, but the further subdivision of the tones caused by the introduction of duplicates of certain keys introduces even more.

For example, when discussing the 18-note keyboard he describes a complete traversal in ascending frequency of all the notes. This is shown below and reveals the elegance of the arrangement. Starting from C_1 we have a small semitone (25/24 in red) to $C\#_2$, then a Great Dieses (128/125 in blue) to $D\flat_3$, and then another small semitone to D_4. From D_4 to D_5 is a syntonic comma (81/80 in green) and that completes the major tone $C_1 D_5$.

The same pattern of red, blue, red then repeats from D_5 to E_{10} giving the minor tone $D_5 E_{10}$. This is then followed by a diatonic semitone (16/15 in orange) from E_{10} to F_{13} which completes the perfect fourth from C to F.

There is then a small semitone (red) to $F\#_{14}$, a syntonic comma (green) to $F\#_{15}$, and then a diatonic semitone (orange) from $F\#_{15}$ to G_{19} which completes the perfect fifth from C to G.

And so it goes on with finally a diatonic semitone (orange) from B_{30} back to C_1 to complete the octave.

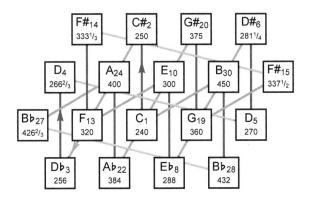

Traversing the schema in ascending frequency.

sml semitone $25/24$	$81/80$	diatonic semitone $16/15$		sml semitone $25/24$	$2048/2025$	$81/80$	ds-m $250/243$	$81/80$
Larger Limma $135/128$		GD $128/125$	sml semitone $25/24$	sml semitone $25/24$	GD $128/125$		sml semitone $25/24$	
diatonic semitone $16/15$			$81/80$	ds-m $250/243$	$81/80$	diatonic semitone $16/15$		sml semitone $25/24$
Great Limma $27/25$				sml semitone $25/24$		Great Limma $27/25$		ds-m $250/243$
major tone $9/8$						minor tone $10/9$		
minor tone $10/9$					$81/80$	minor tone $10/9$		
major third $5/4$								
minor third $6/5$								sml semitone $25/24$

Various subdivisions of a major third.

Further subdivisions occur with the 26-note keyboard where we find that the major tone $F_{13}G_{19}$ is divided thus

$F_{13}F\#_{14}$	$25/24$, small semitone
$F\#_{14}F\#_{15}$	$81/80$, syntonic comma
$F\#_{15}G\flat_{16}$	$2048/2025$, Diaskhisma
$G\flat_{16}G\flat_{17}$	$81/80$, syntonic comma
$G\flat_{17}G_{18}$	$250/243$, demiton souz-minime
$G_{18}G_{19}$	$81/80$, syntonic comma

Note the new small intervals, the Diaskhisma of 19.6 cents and the interval of $250/243$ or 49.2 cents which Mersenne calls the demiton souz-minime.

Various subdivisions of a major third are shown in the diagram above. This reveals that the differences between a minor tone and a major tone, between a diatonic semitone and a Great Limma, between a small semitone and a Larger Limma, and between the demiton souz-minime and the small semitone are all a syntonic comma. Observe also that

a Diaskhisma plus a syntonic comma make a Great Diesis,

a Great Diesis plus a Larger Limma make a Great Limma,

a Great Limma plus a small semitone make a major tone.

Moreover, when discussing Pythagorean scales, we noted that

a Pythag comma plus a Pythag Limma make a Pythag Apotome,

a Pythag Apotome plus a Pythag Limma make a major tone,

two major tones make a Pythag major third.

Gosh! I think that's quite enough of this interval game.

A better keyboard

IT IS INTERESTING to consider how Mersenne's keyboards could be improved. Suppose we wish to support all major scales from five flats to five sharps and that no duplication is allowed (as would happen if C were duplicated in Mersenne's keyboard since C major could then be obtained in two ways).

If we wish to do the scales using the two row schema mentioned earlier then that requires 27 keys. However, it does not allow any minor scales at all. If we extend the schema to three rows to cover minor scales then we end up with a horrendous 41 keys per octave.

By using multiple rows much more efficient arrangements are possible. Thus consider the schema below. The total number of keys is 26 and in the schema they are arranged as two rows of seven and two rows of six. This is very much in the style of Mersenne's 18-note keyboard and is simply a logical extension of it.

This duplicates all white keys except C and E so there is still only one way of doing C major. Note however that both F and A are different from those in Mersenne's keyboard. The extra notes are F at 324 and A at 405 and are a syntonic comma higher than the F and A for C major whereas the other extra white keys are a syntonic comma lower than the corresponding keys for C major.

In the case of the black keys F#, C#, B♭, and E♭ are duplicated (and the second C# has not been met before).

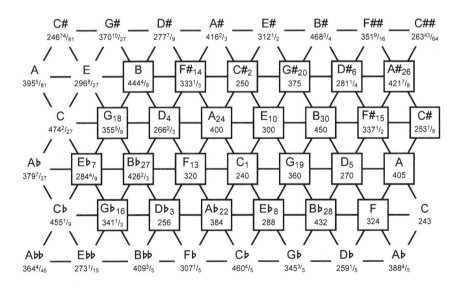

A schema for a 26-note keyboard covering major scales from 5 flats to 5 sharps.

253 1/8 C#		288 Eb₈		337 1/2 F#₁₅			432 Bb₂₈		506 1/4 C#
250 C#₂	256 Db₃	281 1/4 D#₆	284 4/9 Eb₇	333 1/3 F#₁₄	341 1/3 Gb₁₆	375 G#₂₀	384 Ab₂₂	421 7/8 A#₂₆	426 2/3 Bb₂₇	500 C#
	270 D₅			324 F	360 G₁₉		405 A	450 B₃₀	
240 C₁	266 2/3 D₄	300 E₁₀	320 F₁₃	355 5/9 G₁₈	400 A₂₄	444 4/9 B₂₉	480 C₃₂		

Layout for a 26-note keyboard covering all major scales from 5 flats to 5 sharps.

A possible layout of the keys for this 26-note keyboard is shown above. In the case of the duplicated keys, that with higher frequency is at the top. When not duplicated the single key is at the bottom. This means that care is needed when playing C major because it uses the upper D, G, and B but the lower F and A. Maybe that will remind us that CD, FG, and AB are major tones (and so we step diagonally up) whereas DE and GA are minor tones and so we step diagonally down.

In the case of minor scales, it can do from five flats to two sharps in full. To complete the minor scales in full up to five sharps requires D# to F## in the top row to complete 3, 4, and 5 sharps. That would be an additional 5 notes giving 31 in total – the same as Mersenne's keyboard.

The F## (F double sharp, usually written Fx) would be slightly irritating because it would complicate the layout of the keyboard.

The overall arrangement is pleasingly symmetric. It doesn't have weird keys such as Cb and E#. The only duplicated black keys are those most commonly used, and there is a good white space between the groups of black keys so that navigation is easier. It would be a nice keyboard. One wonders why Mersenne did not consider such an arrangement.

As a final remark, note an important difference between the pure two or three row schemas and the more efficient multiple row arrangements. In the case of multiple rows this means that as we shift up or down one sharp or flat, sometimes the pattern just shifts left or right but sometimes it goes up or down a row as well. It changes row going up from G_{19} (1 sharp) to D_4 (2 sharps) and also going down from Bb_{27} (2 flats) to Eb_8 (3 flats). This can have a serious impact on modulation as will now be discussed.

Modulation

WE NOW CONSIDER transitions between keys which reveals yet another problem with just intonation. Transition from one key to another frequently occurs in even short pieces of music. A piece will start in one key and then change to a different key and perhaps finally return to the original key.

Consider for example Chopin's valse, op 64 no 3. It starts in A flat major (four flats), switches to C major (no flats) in the middle, and finally returns to the original key of A flat major. This particular transition is done via G in the treble which is the leading note in A flat major (the note that leads into the tonic) and becomes the dominant in C major (a fifth from C). In our keyboard the common G is G_{19} and the transition is smooth. However, if we had been using the two row arrangement then the G in Ab major would be G_{18} whereas the G in C major would be G_{19}. So an awkward discrepancy of a syntonic comma would arise.

As another example consider op 18 no 1. This valse starts in E flat major (three flats), switches to D flat major (five flats) and then returns to E flat major. The transition in this case is via A♭ which is the subdominant (a fourth above the tonic) in the first key and becomes the supertonic in the second. Again these are both $A♭_{22}$ and so the transition is smooth. Suppose however that the whole piece were transposed to start in F (one flat). It would then modulate to E♭ (three flats). The transition note would now be B♭ but this time we have a problem because in F the appropriate B♭ is $B♭_{27}$ whereas in E♭, it becomes $B♭_{28}$. So again there is a discrepancy of a syntonic comma.

An interesting example is the polonaise op 61. This starts in A flat major (four flats) and then switches to E major (four sharps). The transition seems to be via A♭ which is the tonic of the first key and transforms into G# which is the mediant (a major third above the tonic) for the second key. In an equitempered keyboard these would be the same and no problem would arise. But in our 26-note keyboard we have to sneakily slip from $A♭_{22}$ to $G\#_{20}$. But these are a Great Diesis of 41 cents apart. So presumably it would sound nasty.

It is almost time to draw this discussion to a close. One could go mad in trying to understand the best way to provide many just scales. Just intonation does not work with a simple keyboard and gives rough intervals if one deviates far from C major. The use of giant keyboards was studied for a long time. But they were expensive to build and hard to play. So it is not surprising that the equitempered scale won the day. We conclude by revisiting the Tonnetz lattice.

Tonnetz revisited

THE TONNETZ LATTICE can be easily adopted for equitempered tuning in which case the lattice repeats in a very simple manner. Thus all Cs are exactly the same rather than differing by multiples of a syntonic comma. Similarly all Ds are exactly the same and an equitempered tone above C. In addition G# is the same

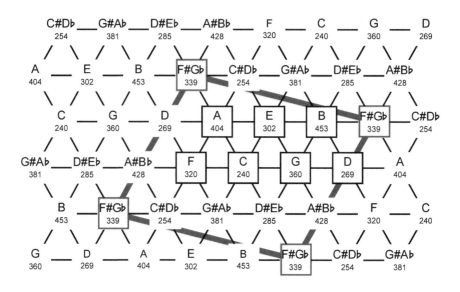

as A♭, F# is the same as G♭, and so on; and the doubly chromatic notes such as F## just become G. The resulting pattern then repeats in both directions as shown above where the repeated unit is a parallelogram. We can choose a note for the vertices of the parallelogram. In the schema above we have chosen to use F#G♭ (in red) at the vertices largely because this shows the scale of C major (in black) within the unit parallelogram. The parallelogram is outlined with bold red lines in the background. Note that D is actually on the boundary. The frequencies are given to the nearest integer.

A parallelogram with opposite sides identified is a representation of a torus (a torus is the shape of a rubber ring which in turn is simply a cylinder bent around). As an example of a map on a torus, we can take part of the uniform hexagonal tiling of a plane by hexagons. This gives a map of seven hexagons on a torus where each one is surrounded by all the others as illustrated below.

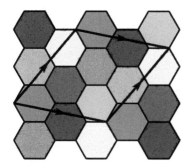

The hexagonal tiling of the plane showing the derivation of the torus.

A torus with a map of seven hexagons; each one is surrounded by the others.

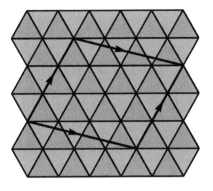 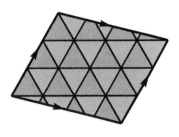

A map of 24 triangles on a torus.

The plane can also be uniformly tiled by triangles. And again we can use this tiling to show how various triangles can be mapped onto a torus. The diagram above shows the mapping of 24 triangles with just two colours. The map has 12 vertices.

It follows that the equitempered Tonnetz schema is essentially a map of 24 triangles on a torus where each triangle represents a major or minor triad and each vertex represents a note. The resulting schema is shown below where the major triads are in red and the minor triads are in blue. The 12 notes are shown in circles. Remember that the black notes are shared and so are indicated by the pairs G#Ab, C#Db, and so on. In order to avoid clutter in future diagrams, we usually just show the more common use, such as F#.

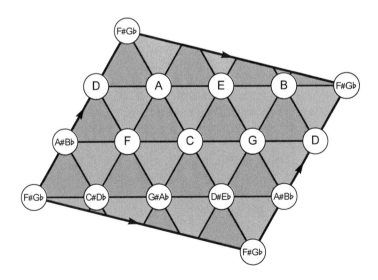

The equitempered Tonnetz schema on a torus.

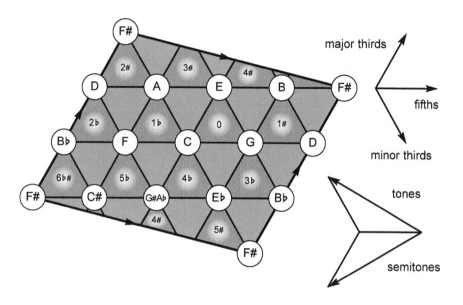

This Tonnetz schema can been used to illustrate many facts. Consider first the progression of fifths, namely – C, G, D, A, E, B, F#, C#, G#A♭, E♭, B♭, F, C. This visits every note and forms a closed spiral going around the torus three times. It follows the horizontal lines in the diagram above.

The progression of major thirds from C is C, E, G#A♭, C on the right-sloping diagonal and goes around the torus just once; there are four such progressions. The progression of minor thirds from C is C, E♭, F#, A, C on the left-sloping diagonal and also goes around the torus once; there are three such progressions.

The progression of semitones starting from C crosses a pair of triads to C# sloping downwards to the left and visits every note and wraps around the torus four times. The progression of whole tones from C crosses a pair of triads sloping upwards to the left and visits half of the notes and wraps around the torus twice; there is another progression of tones starting from B which visits the other notes.

Each major triad (red triangle) such as CEG defines the root of a major scale; it is also known as the root chord of that scale which, in this case, is C major. In the diagram above the red triangles are labelled indicating the number of sharps or flats in each scale. Note that six sharps is the same as six flats in the equitempered world.

Similarly, each minor triad (blue triangle) such as CE♭G defines a minor scale which, in this case, is C minor. It has the same key signature as the major scale to its immediate right, namely E flat major with 3 flats.

Each major triad such as CEG is surrounded by three minor triads and so each major scale is associated with three minor scales. We have already met the association of C major with C minor, triad CE♭G, three flats; the triads share the perfect fifth CG. We have also met the association with A minor, triad ACE, no sharps/flats; the triads share the major third CE. The third and new association is with E minor, triad EGB, one sharp; the triads share the minor third EG.

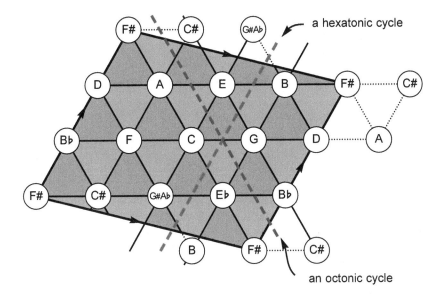

a hexatonic cycle

an octonic cycle

Such adjacent triads having two notes in common are of two kinds. If the triads share a major third such as CE, then the unshared notes differ by a tone (in this example, A versus G). In the other cases the unshared notes are a semitone apart (B versus C and E versus E♭).

Observe that scales which might seem at first to be far apart can be quite close. Thus C major is close to A flat major (four flats) and this transition is one which was mentioned earlier when discussing modulation (Chopin valse op 64 no 3). Similarly, four flats and four sharps are close (Chopin polonaise op 61).

The diagonal strip with sides BGE♭ and ECA♭ forms a sequence of six triads which wrap around the torus. This sequence is known as a hexatonic cycle and such cycles are found in many classical works. An example is in Brahms' Double concerto (first movement). Clearly there are four such hexatonic cycles. The reader should try them out on a handy piano. Transition along the cycle is smooth with only a semitone difference at each step.

We can make a similar set of transitions along the other diagonal sequence thus: CEA, EAC#, AC#F#, C#F#B♭, F#B♭E♭, B♭E♭G, E♭GC, GCE, and back to CEA. This is a sequence of eight triads where four of the transitions have a whole tone difference. Perhaps not so smooth. We can refer to it as an octonic cycle. There are three such cycles. An example is in Schubert's overture *Die Zauberharfe*.

The third direction following the sides of the triangles (that is horizontal in the diagram) traverses the scales in numerical order. If we start from C#F#A and progress left to DAF we find part of Brahms' symphony number 2.

Another obvious group of related triads is those around a central note. Thus around G clockwise we find GEB, GBD, GDB♭, GB♭E♭, GE♭C, GCE, and back to GEB. Again half of the transitions involve a tone difference. An example around A is found in Verdi's *Il Trovatore*.

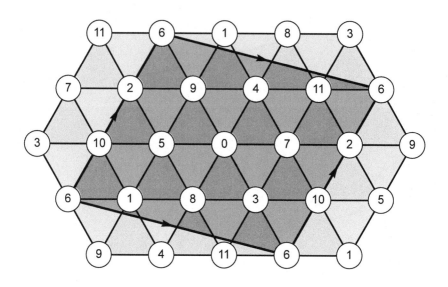

The schema using numbers modulo 12.

It is clear that it is the triads that matter and not so much the major and minor scales themselves. Thus it seems to justify the thought that Mersenne's reason for his choice of notes in his smaller keyboards was primarily driven by the desire to maximize the possible triads.

It is an interesting exercise to draw the schema but with numbers replacing the notes. In the extended diagram above C has been replaced by 0, C# by 1 and so on. The corners of the parallelogram (F#) become 6 and B becomes 11. It is then clear that going around the torus is all done with arithmetic modulo 12. Moving NE adds 4, moving SE adds 3, and moving W adds 5. These add to 12 of course since we have simply gone around a triangle.

Another thought is to consider Euler's formula for maps. If a map on a sphere has F faces, E edges, and V vertices, then $F+V-E$ is always 2. We say that the Euler number for a sphere is 2. As an example, a cube has 6 faces, 12 edges, and 8 vertices. And indeed $6+8-12 = 2$.

On a torus the Euler number is not 2 but 0. And indeed in our Tonnetz map we have 24 triangles, 12 vertices, and 36 edges. And $24+12-36 = 0$. The 24 triangles correspond to 12 major triads plus 12 minor triads. The 36 edges correspond to 12 perfect fifths plus 12 major thirds plus 12 minor thirds. The 12 vertices correspond to the 12 notes of the equitempered scale.

It is an important fact that there are exactly 12 semitones in the octave. And that 12 is an abundant number and in particular is divisible by 3 and by 4. As a consequence, three major thirds and four minor thirds both make an octave. If this were not the case then the hexatonic cycle would not exist and the harmonious possibilities would be absent. But the real miracle is that with the equitempered scale divided by 12 nevertheless the various intervals such as the perfect fifth are tolerable approximations to the just intervals. The 53-note equitempered scale which we encountered in the lecture on Music would be hopeless in the broader sense because 53 is a prime number and the octave would not subdivide into an exact number of major and minor thirds (or indeed into an exact number of anything) and the harmonious cycles would not exist.

It might be wise to stop there and let the reader browse at will and tickle harmoniously at that old piano. But let us draw this book to a close by ruminating over the fact that we seem to have discovered that 12 really has to be our favourite number. If only we had 12 fingers!

Further reading

THIS TOPIC has fascinated mathematicians and musicologists for several centuries. For a good description of Tonnetz and its background see the Wikipedia article at https://en.wikipedia.org/wiki/Tonnetz.

An interesting book which covers many aspects of this fascinating topic and its use in the analysis of music is *Audacious Euphony: Chromatic Harmony and the Triad's Second Nature* by Richard Cohn. The examples briefly mentioned in the final section above are described in detail in this book.

And dedicated readers should brush up their Latin and consult *Tentamen novae theoriae musicae ex certissismis harmoniae principiis dilucide expositae* by Euler. And also activate their old French and read *Harmonie Universelle* by Mersenne.

For a description of the maps on a torus and other topological figures see *Gems of Geometry* by the author.

Bibliography

The following are referred to in the text. Note that the edition given in some cases may not be the latest but is simply the one in my library. There is of course an enormous amount of material on the web as well.

Barbour, J Murray, *Tuning and Temperament*, Michigan State College (1951)

Barnes, John G P, *Gems of Geometry*, 2nd edition, Springer (2012)

Barnes, John G P, *Programming in Ada 2012*, CUP (2014)

Burton, David M, *Elementary Number Theory*, Allyn and Bacon (1980)

Cohn, R, *Audacious Euphony: Chromatic Harmony and the Triad's Second Nature*, OUP (2012)

Colenso, John William, *Arithmetic Designed for the Use of Schools: To which is Added a Chapter on Decimal Coinage*, Longmans, Green (1886), reproduced by Bibliolife

Conway, John H and Guy, Richard K, *The Book of Numbers*, Copernicus (1996)

Craik, Alex D D, *Mr Hopkins' Men*, Springer (2007)

Coxeter, H S M, *Introduction to Geometry*, 2nd edition, Wiley (1969)

Davenport, H, *The Higher Arithmetic*, 7th edition, CUP (1999)

Dudeney, H E, *Amusements in Mathematics*, Dover (1970)

Duncan, David E, *The Calendar*, Fourth Estate (1998)

Fauvel, John, Flood, Raymond, and Wilson, Robin (eds), *Music and Mathematics*, OUP (2003)

Feller, William, *An Introduction to Probability Theory and Its Applications*, Wiley (1957)

Gibbons Simplified Stamp Catalogue, Stanley Gibbons Ltd, (1952)

Gillings, Richard J, *Mathematics in the Time of the Pharaohs*, Dover (1982)

Grossman, Israel and Magnus, Wilhelm, *Groups and their Graphs*, The Mathematical Association of America (1964)

Heath, Thomas L, *The Thirteen Books of Euclid's Elements*, Dover (1956)

Helmholtz, Hermann, *On the Sensations of Tone*, Dover (1954)

Hollingdale, S, *Makers of Mathematics*, Penguin (1994)

Johnston, Ron, *Bell-Ringing: The English Art of Change-Ringing*, Viking Books (1986)

Kaplan, Robert, *The Nothing That Is*, Allen Lane (1999)

Knuth, Donald E, *The Art of Computer Programming, Vol 1*, Addison-Wesley (1968)

Lindley, Mark, *Lutes, Viols and Temperaments*, CUP (1984)

Nahin, Paul J, *An Imaginary Tale – The Story of $\sqrt{-1}$*, Princeton University Press (1998)

Newman, James R, *The World of Mathematics*, Simon and Schuster (1956)

Packel, Edward, *The Mathematics of Games and Gambling*, The Mathematical Association of America (2006)

Riesel, Hans, *Prime Numbers and Computer Methods for Factorization*, Birkhäuser (1985)

Rosen, Kenneth H, *Elementary Number Theory and its Applications*, Addison-Wesley (1993)

Rouse Ball, W W, *Mathematical Recreations and Essays*, 11th edition, Macmillan (1939)

Rubik, Ernö et al, *Rubik's Cubic Compendium*, OUP (1987)

Semple, Clara, *A Silver Legend, the Story of the Maria Theresa Thaler*, Barzan (2007)

Singh, Simon, *The Code Book*, Fourth Estate (2000)

Singmaster, David, *Notes on Rubik's 'Magic Cube'*, Singmaster (1980)

Snowdon, Jaspar and Snowdon, William, *Diagrams*, Christopher Groome (1972)

Stedman, Fabian, *Campanologia*, (1677), reproduced by Christopher Groome (1990)

Stedman, Fabian, *Tintinnalogia or The Art of Ringing*, WG (1668), reproduced by Kingsmead reprints (1970)

Stewart, Ian, *Nature's Numbers*, Basic Books (1995)

Wells, David, *The Penguin Dictionary of Curious and Interesting Numbers*, Penguin (1997)

Yan, Song Y, *Perfect, Amicable and Sociable Numbers*, World Scientific (1996)

Index

Abel   249, 266
abundant numbers   5
Ackermann   228
alternating group   256
amicable pair   35
angle notation   121
apothecaries' measure   12
approximations   92, 201
Arabic numerals   115
Argand plane   170, 174
Augustus   103
avoirdupois weight   11

Babylonian numbers   120
bases   8, 121, 247
Beast   2, 119
bells   137, 254
Berlin papyrus   87
binomial coefficients   45, 231
binomial distribution   45, 52
bob   149, 159
Bosanquet   203
Buffon   57

Caesar   102
calendar   101
camels   22, 116, 218
Carmichael numbers   134
Cauchy distribution   55
Cayley   251
cents   183
chain measure   13
change   138
Chapman   241
Chinese Remainder Theorem   295
 eggs   300
 pirates   299
 squares   301
Chinese Rings   222, 287
Chopin   317
Cole   34
Colenso   20, 23
commutation   139
complex numbers   169
 conjugate   171
 primes   173
congruency   28
 linear   209, 297

continued fractions   90, 201
 square roots   91
convergents   92
coordinates   172
cosets   159, 252, 280
craps   61
cryptography   205
cube roots   89
currencies   19
cyclic group   249
cycle length   83, 129, 131

decimalization   20
decryption   206, 214
de Moivre   293
denominator   71
derangement   292
Descartes   172
dice   49
Diophantus   209, 295
difference equations   287
digital root   125
dihedral group   250, 254
dodging   141, 151
doubles   139
distributions
 binomial   45, 52
 Cauchy   55
 normal   53
divisibility   124, 245
Dozenal society   17
Dudeney's puzzles
 barrels   125
 cannonballs   236
 hats   290

Egyptian
 fractions   71
 numerals   116
 squares   88
 weights and measures   94
ellipse   100
encryption   206, 214
equal temperament   196
equation of time   99
equinox   98
Eratosthenes   165, 168, 174

estimating 48
  pi (π) 56
Euclid 4, 26, 162
  algorithm 162, 209, 295
Euler 5, 27
  function 212
  theorem 212
  Tonnetz schema 306

factorial 6, 227
factorization 165, 166, 176
favourite numbers 1
  seven 2, 95
  twelve 3, 5, 323
  thirty-seven 3, 84, 86
Fermat 40
  factorization 166
  numbers 40
  Last Theorem 40
  Little Theorem 133, 213, 216
Fibonacci numbers 41, 77, 94, 234
  and greatest common divisor 163
  explicit formula 289, 290
four group 261

game theory 63, 66
gaits 217
Gauss 232
golden number 41, 77, 94
grandsire method 157
greatest common divisor 161, 209
Gregorian calendar 103
groups 159, 249, 266, 278
Gunter 13

Hamilton 262
Handel 188
Henry I 16
hexatonic cycle 321
histogram 44
Hopkins 20, 251
horses 219
Horus 94
hunting 138

irrational numbers 69
isomorphic groups 255
iteration 228

Julian calendar 102
just intonation 185

KDF9 computer 80, 230
Kepler's laws 98

Kolmogorov 241

Lagrange 17, 260
Laplace 17
Law of Chance 43, 54
Law of Large Numbers 48, 54, 56
leap second 101
leap year 101, 102
Lehmer 30
Leonardo of Pisa 41
likelihood 48, 61
linear congruence 209, 297
logarithms 53, 93, 172
Lucas 30

major sixth 181, 305
major third 181, 185, 305
major tone 181
mean 47
meantone temperament 193
medical statistics 58
Mersenne 29, 131, 188, 303
minor sixth 190, 305
minor third 190, 305
minor tone 185
modular arithmetic 28, 124, 176
  inverse 29, 206, 208
modulation 317
multiplication table 123
multiplicative functions 26, 213
Montmort 292
musical notes 18

natural logarithms 53, 93, 172
normal distribution 53
numbers
  complex 169
  irrational 69
  transcendental 69
numerator 75

octonic cycle 321
orbit of Earth 100
order of group 250, 260

Paganini 36
Pascal 45
  triangle 45, 231
pentagon 41
perfect fifth 180, 305
perfect fourth 180, 305
perfect numbers 5, 25, 30
permutations 139, 254
  on Rubik cube 278

Index

pi (π) 56
place notation 115
plain course 148, 159
plain hunting 138
poker 63
polygon 41
polynomials 176
Poole 203
prime numbers 3, 161, 173
  complex 173
  in cryptography 205
  largest 34
  Mersenne primes 29
Poisson 242
Poulet 39
pyramids 235
Pythagorean scale 181
Pythagorean triples 237

quaternions 262
queuing 242
Qurra 37

real projective plane 143
recurring fractions 82, 129
  continued fractions 91
  cycle length 83, 129, 131
recursion 220, 227
Reductio ad Absurdum 34, 70
Rhind papyrus 71
Roman numbers 117
Rubik cube 265

St Ives 96
Salmon 200
semitone
  chromatic 182
  diatonic 182
  equitempered 197
  just 185
seven 2, 95
seventh 201, 203
sidereal day 97
sieve of Eratosthenes 165, 168, 174
smoothness 36
sociable cycles 39

solar day 97
solstice 98
square pyramids 235
square roots 87
  complex 172
  continued fractions 91
standard deviation 47
statistics 48, 61
  medical 58
Stedman 151
Stirling 54
stochastic processes 66, 239
subfactorial 292
subgroups 250
sundial 99
sunshine 106
superabundant numbers 5
Sylow 260
syntonic comma 86, 185

tax year 105
Thabit ibn Qurra 38
thirty-seven 3, 84, 86
twelve 3, 5, 323
tilt of Earth 98, 106, 111
time notation 121
Tonnetz schema 303, 306, 317
  on torus 319
torus 318
touch 153
Tower of Hanoi 220, 288
transcendental numbers 69
triad 192, 306, 319
triangular numbers 232
triangular pyramids 232, 235
triple 156
troy weight 11

unrelated events 43

variance 47

Wiles 40
wolves 194
work 141
wrangler 21